Skywriting
An Aviation Anthology

Skywriting
An Aviation Anthology

JAMES GILBERT

ST. MARTIN'S PRESS
NEW YORK

Contents

Introduction

Though this book follows a rough time order, it is in no way a history of aviation, but simply a collection of excerpts from those books about flying which have given me the greatest pleasure. It springs from half a lifetime spent partly as a bookworm and partly flying; and from being hired by Arno Press Inc., a subsidiary of *New York Times*, to select a series of aviation books that were out of print but which we felt deserved reprinting for libraries and other collectors. (This 36-volume Arno collection is entitled *Literature and History of Aviation*.) The year I spent on this project filled my head with ideas for an aviation anthology: here it is.

I believe there is something of almost every aspect of aviation. First the balloonists, riding their imprisoned bubbles of gas, the first (and for 120 years, the only) men to see the earth from above. Then the pioneers of winged flight, wrestling with a medium that was insubstantial, unknown, invisible. Then, just a decade after their first successes, world war, and the frenzied development of the aeroplane as a weapon. Two decades of uncertain peace followed, and the slower evolution of the aeroplane as a vehicle, both public and private, and as a means of navigation by adventurers across the oceans. (Here grew the roots of the airlines.) World War II came next, and the aeroplane was employed as a weapon first to halt the advance of tyranny, and then as a bludgeon to batter that same tyranny into dreadful ruins.

Since then only a few lesser wars; and instead, the extraordinary growth of the airlines, and of private flying. The last excerpt quoted is from an account of man's greatest feat of exploration: the voyage to the moon.

Acknowledgments

The compiler is indebted to the following for permission to quote the extracts that comprise this anthology:

Curtis Brown Ltd and the estate of C. C. Turner for the excerpts from *The Old Flying Days* by C. C. Turner.

Curtis Brown Ltd for the excerpt from *Sky Fever* by Sir Geoffrey de Havilland.

Peter Davies Ltd and The K. S. Giniger Co. Inc. and Stackpole Books Inc. for the excerpt from *Sagittarius Rising* by Cecil Lewis, © 1936 Cecil Lewis.

Doubleday & Co. Inc. for the excerpt from *Ace of the Iron Cross* by Stanley M. Ulanoff. Copyright © 1970 by Stanley M. Ulanoff.

Holt, Rinehart & Winston Inc. and Routledge & Kegan Paul, Ltd for the excerpt from *Flying Dutchman* by Anthony Fokker. Copyright 1931, © 1959 by Bruce Gould.

Prentice-Hall Inc., Englewood Cliffs, New Jersey, for the excerpt from *Rickenbacker: an autobiography*, © 1967 by Edward V. Rickenbacker.

Simon & Schuster Inc. for the excerpts from *Old Soggy No. 1* by Hart Stilwell and Slats Rogers.

Charles Scribner's Sons Inc. and John Murray (Publishers) Ltd for the extract from *The Spirit of St Louis* by Charles A. Lindbergh. Copyright 1953 Charles Lindbergh.

The estate of the late Nevil Shute, William Heinemann Ltd and William Morrow & Co. Inc. for the extracts from *Slide Rule* by Nevil Shute.

Hodder & Stoughton Ltd and Coward, McCann & Geoghegan Inc. for the extract from *The Lonely Sea and the Sky* by Francis Chichester.

The Ziff Davis Publishing Company Inc. and Hodder & Stoughton Ltd for the extract from Ernest K. Gann's *Flying Circus*.

MacDonald and Jane's Publishers Ltd for the extract from *On Being a Bird* by Philip Wills.

Faber and Faber Ltd and Random House Inc. for 'Airman's Alphabet' from *The Orators* by W. H. Auden.

Chatto and Windus Ltd and David Garnett for the extract from *A Rabbit in the Air* by David Garnett.

Harcourt Brace Jovanovich Inc. and William Heinemann Ltd for the extract from *Wind, Sand and Stars* by Antoine de Saint-Exupéry, translated by Lewis Galantiere, © 1939 by Antoine de Saint-Exupéry, © 1967 by Lewis Galantiere.

Country Life and Harald Penrose for the excerpt from *I Flew With the Birds* by Harald Penrose.

Hodder & Stoughton Ltd for the excerpt from *Fate is the Hunter* by Ernest K. Gann.

Mr Lovat Dickson for the excerpt from *The Last Enemy* by Richard Hillary.

Pierre Clostermann and Chatto and Windus Ltd for the extract from *Flames in the Sky* and *The Big Show* by Pierre Clostermann.

William Kimber & Co. Ltd for the extract from *Duel Under the Stars* by Wilhelm Johnen.

Paul R. Reynolds Inc., 599 Fifth Avenue, New York, N.Y. 10017 for the extract from *Mission with LeMay*, © 1965 by Curtis LeMay with MacKinlay Kantor.

W. W. Norton & Co. Inc. for the extract from *Serenade to the Big Bird* by Bert Stiles. © 1947 by Mrs Bert W. Stiles, copyright renewed 1973.

Colonel Gregory Boyington for the extract from *Baa Baa Black Sheep* by Col. Boyington.

Evans Brothers Ltd for the extract from *I Flew for the Führer* by Heinz Knoke.

Houghton Mifflin Co. Inc. and Mr Guy Murchie for the extract from *Song of the Sky* by Guy Murchie.

Van Nostrand Inc. for the extract from *Air Power, the Decisive Force in Korea* by Colonel Harrison R. Thyng.

Cassell & Collier Macmillan Publishers Ltd for the extract from *Stranger to the Ground* by Richard Bach.

Cassell & Collier Macmillan Publishers Ltd for the extract from *No Echo in the Sky* by Harald Penrose.

Air Force Magazine, published by the Air Force Association, 1750 Pennsylvania Ave., N.W., Washington, D.C. 20006, for 'Night Mission on the Ho Chi Minh Trail' by Mark E. Berent.

Harry N. Abrams Inc., 110 East 59th Street, New York, N.Y. 10022, for the extract from *Always Another Dawn* by A. Scott Crossfield, with Clay Blair, Jr.

W. H. Allen & Co. Ltd and Farrar, Straus & Giroux Inc. for the extract from *Carrying the Fire* by Michael Collins, with a foreword by Charles A. Lindbergh. © 1974 by Michael Collins; foreword © 1974 by Charles A. Lindbergh.

Thanks are also due to the copyright owners of the following items, whom the compiler has been unable to trace:

Aircraft by Le Corbusier.

The Dying Aviator and *In Other Words*.

Dancing the Skies by John Gillespie Magee.

The Blimp That Came Home—Without Its Crew by Jack Pearl.

The airplane takes possession of the sky — the various skies of Earth.

The airplane, symbol of the New Age.

It is high enough, up there in the sky. You must lift your head to suit.

Lift head and look above.

The airplane, advance guard of the conquering armies of the New Age, the airplane arouses our energies and our faith.

Le Corbusier, *Aircraft*, 1935.

Men have been sailing the skies for almost two hundred years; but only for the last eighty in aeroplanes. For one hundred and twenty years, ballooning was the only way man could ascend into the heavens. The first balloons were constructed in the 1780s by two French brothers Joseph and Jacques Étienne Montgolfier who were paper manufacturers; from observing that hot air rises, they went on to discover that air heated to 180°F loses half its weight (volume for volume), and that light paper bags filled with hot smoke would lift.

They proceeded most gradually, from private experiments to public demonstrations of tethered aerostats with no one aboard; then to demonstrations before King Louis XVI, ever prepared to have his crowning boredom distracted by new amusements. One day at Versailles, the Montgolfiers were allowed to launch a balloon bearing a sheep, a rooster and a duck. Two men made a tethered ascent, but the king would not allow a free manned flight, except: 'Take two criminals who are under the death sentence and tie them to the basket of the balloon. That will be a novel way of being rid of worthless men!' And his courtiers tittered. But the Marquis François-Laurent d'Arlandes, full of youthful and aristocratic boldness, pleaded: 'Your Majesty! The glorious honour of being the first to fly should not go to two such vile villains! Only worthy subjects of your glorious kingdom are fit to be the first to soar into God's heavens. I beg you to allow the physician de Rozier and I this honour.' What could Louis say but yes? The two adventurers' first flight was an ambitious one; they climbed to 3,000 feet, and were airborne for twenty-five minutes.

The Montgolfiers wrongly thought that burning their chopped straw and wool fuel released some special lightweight gas unknown to science, and that from this sprang their buoyancy. The lightest gas, hydrogen, had already been discovered, and the first manned flights in a hydrogen balloon followed just ten days later.

Ballooning has always been practically useless, very expensive, and prodigious slow fun. The perfection of propane gas burner systems and synthetic fabrics for the envelope have lately brought the costs and hazards of hot air ballooning down to a level allowing a vast mushrooming of it as a sport. There are still a few rich eccentrics ballooning with hydrogen-filled envelopes.

The nicest descriptions of the joys of gas ballooning I know are by the pioneer English aeronautical journalist Major C. C. Turner. Here are his accounts from Edwardian times of a solo flight by day, and

13

part of a long night voyage (actually an attempt on the world record for distance) made with two companions.

1

FROM: *The Old Flying Days*
BY   C. C. Turner
Sampson Low & Marston, London, 1927

By day

I am alone in a very small balloon over Surrey on the last day of a beautiful May. So light is the wind that on that little voyage three hours and a half were taken to travel sixteen miles, from Roehampton to Betchworth. There is scarcely a cloud in the sky. The sun is hot, and the balloon, being one of gold beaters' skin inflated with hydrogen, is very 'lively'. Hydrogen is much more sensitive to heat and cold than coal gas, and the goldbeaters' skin is not so good a protection as ordinary balloon fabric. The basket is quite tiny; two men in it would be a crowd! The rigging and all the appointments are designed for lightness. I am alone, and as the hours pass the loneliness becomes rather oppressive. This is quite different from aeroplaning, for there is little to occupy the attention, and scarcely anything to do with the hands. Plenty of time to look around and down!

Very slowly I approach a big wood. It would better express the situation were I to say that very slowly a big wood comes nearer to the balloon, for there is no sense of movement, and the earth below seems to be moving slowly past a stationary balloon. As the wood comes nearer I watch the aneroid and get a bag of ballast ready, for I know that over the wood the air will be slightly cooler, with probably a slight down draught due to convection, and that the balloon will immediately begin to descend; and I shall have to check that descent by throwing a little ballast.

Fifteen hundred feet up and almost absolute silence, broken occasionally by the barking of a dog heard very faintly, or by a voice hailing the balloon, and by an occasional friendly creak of the basket and rigging if I move ever so slightly. Then quite suddenly I am aware of something new.

The balloon has come down a little already, and I scatter a few

14

handfuls of sand and await the certain result. But my attention is no longer on that, it is arrested by this new sound which I hear, surely the most wonderful and the sweetest sound heard by mortal ears. It is the combined singing of thousands of birds, of half the kinds which make the English spring so lovely. I do not hear one above the others; all are blended together in a wonderful harmony without change of pitch or tone, yet never wearying the ear. By very close attention I seem to be able at times to pick out an individual song. No doubt at all there are wrens, and chaffinches, and blackbirds, and thrushes, hedge sparrows, warblers, greenfinches, and bullfinches and a score of others, by the hundred; and their singing comes up to me from that ten-acre wood in one sweet volume of heavenly music. There are people who like jazz!

But the ballast has taken effect, and the balloon is steadily rising; the music becomes fainter; the wood slips away slowly towards the north.

The balloon steadily mounts to 7,000 feet, for the sun is very hot. Awaking as it seems from a beautiful dream I begin to think of a place to land. There are the southern slopes of the Surrey hills. I pick out Reigate, Betchworth, Dorking. Far way to the south I see the South Downs and the gleam of the Channel. But with the end of my ballast approaching I must prepare to land. My course will take me right over Betchworth, and there is a railway station which will save me a lot of trouble. I valve, and the balloon promptly begins to fall.

Rather a heavy landing, and I am down in the basket huddled up by the fall. Then it bounces, and comes down again almost on the same spot. I hang on to the valve until the balloon is empty enough to permit me to get out carefully as it falls slowly over and the people of the village assemble and offer help.

By night

We left towards the north-east and watched the upturned faces of the crowd sinking away below us, and the Crystal Palace itself, already lit up for the night, receding towards the south-west.

Within a few minutes of leaving the ground we were rewarded with one of the finest spectacles ever seen by the airman. London's 150 square miles were spread westwards before our eyes as we crossed the Thames not far from Greenwich, and gazed at the vast panorama. A crescent moon was not powerful enough to dim the stars, and we seemed to be poised in the centre of a vast illuminated globe whose dark sides were frosted with silver and gold, the roof glittering with the constellations seen, at our height of 2,000 feet, as they never appear to the eyes of the Londoner. Below us lay the millions of lamps patterning the great city, the wide, well-lit highways, such as Oxford Street, conspicuous, and the dark band of the river braceleted by the lights of the bridges. The roar of the traffic came up to us, an endless murmuring, and the whistling of trains and the barking of dogs came clearly to our ears.

Soon we were over the quiet dark country, and with nothing to guide the eye had frequently to read our instruments in the electric light with which we were equipped. But coming to high ground on one occasion we were seen, and a voice hailed us from below; and this enabled us to ascertain our position with more exactness than by studying the map. During the next two hours we slowly ascended to about 4,000 feet, and the balloon was in perfect equilibrium, and needed neither valving nor ballasting. But the question whether we were to risk the sea-crossing became an urgent one. We were steadily approaching the coast, and for a long time had seen the occulting lights about Harwich and northwards of that port. We had to make a decision against time, and the alternative to a landing almost immediately was a crossing over the North Sea on an apparently straight north-east course, probably three or four hundred miles of water if the wind held, whilst if it went round ever so slightly to the south there was a possibility of drifting right up between Scotland and Norway.

Rapidly we weighed the matter up. The balloon was good for two days and nights, for we carried a ton of ballast and disposable material. The weather indications had not suggested any likelihood of change, and we always had the consoling thought that at a height of 17,000 feet or so a westerly stream of air is certain, so that, provided we kept strict watch on the time and did not leave decision until too late, we could fall back upon, or rather rise up to, the safety which that would afford. The one difficulty lay in the fact that for hours to come we should not be able to ascertain the direction in which we were travelling, for when over the sea and out of sight of fixed landmarks it is impossible in a balloon to know the direction of movement. There was just the faint chance that by dawn we might see a ship and be able to take a line on its white wake. Well, we agreed to go on, and we crossed the coast near Yarmouth, knowing we were in for a long night vigil over the sea.

It was cold, and although we were warmly clad the inaction one is forced to in a balloon, except at the rare moments of some crisis, keeps the circulation at a low ebb. We had some cold supper, with hot drink from thermos flasks. Before we left the coast, and shortly afterwards, Gaudron provided us with a novel spectacle. He had brought some coloured flares, and lowering these to the end of our trail rope he lighted them by electric current, so that first green, and then red and then white flares burst out below us, wasted their rays in the emptiness around, but lit up the balloon itself so that the great globe above us and every rope was dazzlingly illuminated. Then, as the fires died out the glorious stars were seen in multitudes and brilliance only seen in favourable conditions from mountain top or in dry climates. The sky was cloudless. Complete silence reigned save for the friendly occasional creak of a rope or the basket as one of us changed position.

The stars seemed friendly and close; Orion and the Great Bear majestic groups, the Pleiades a heavenly cluster, the Milky Way a celestial highway. But about one o'clock a sudden change occurred. At a great distance from us, but at about the same level, a great number of small fluffy clouds formed, and we were poised in the middle of the circle. Another ring formed nearer to us, and another. And we moved on, but with no perceptible sense of movement, and with this weird escort. Once or twice far below us we could see the light of a ship. After two o'clock we fired another flare.

Now we witnessed the formation of cloud on a larger scale, more particularly to the north, and lower than the balloon, and gradually it extended until cloudland blotted out the sea entirely except for an occasional patch of blackness in the grey. Above, the sky was clear, but to the south-west the stars were blotted out by a wall of black cloud. Soon after four o'clock a faint flush appeared in the north-east, marking the approach of the sun to his rising. The water in the wet bulb thermometer was frozen.

At five o'clock the light was strong enough to make a faint shadow. We were at 4,500 feet altitude. The cloud scenery now began to bestir itself. As if for our sole benefit it began a long series of wonderful groupings. Across the north-east sky a straight row of fantastic shapes appeared black as ink against the lightening sky. They resembled gigantic trees rearing themselves from a flat land covered with white mist. These grotesque shapes appeared to be the same clouds that half an hour before had passed slowly below us, then appearing indefinite and fleecy.

The dawn grew nearer, and a red tinge appeared behind the row of cloud trees, which became blacker and more sharply defined. A lovely green hue suffused the sky above the red. To the south the clouds were bluish grey. The stars were still very brilliant.

Almost suddenly the row of strange tree shapes lifted to a higher level, or we sank; then imperceptibly they dispersed and a series of mysterious and ever-changing cloud forms took their place. One slate-grey, ponderous-looking mass occupied a giant's share of the northern sky slightly below us, but with its topmost peaks and domes about on our level. It was tremendous; and I find it is quite impossible to give any idea of the immensity and variety of these changing scenes. In the far south a limitless stretch of cloud peaks looked like Switzerland moulded in snow.

They continued across the North Sea and Denmark to make a rough but safe landing in Sweden, nineteen hours after their take-off. The distance they had flown was 703 miles (including 360 over the sea) – not enough to beat the world distance record, which then stood at 1,197 miles, but still a fine flight.

Washington H. Donaldson was an aeronut and exhibitionist who early discovered (as cats discover mice) that balloons drew crowds, and that there was fame and money in such crowds. He'd been a magician, ventriloquist, tightrope walker and acrobat, and was the first man in the world to ride a bicycle on a high wire. After 1,300 performances of this sort of thing, he tried ballooning and reported, 'It was so glorious that I resolved to abandon the tightrope for ever.' Thereafter he barnstormed America with his balloon, often attaching to it not a basket but just a trapeze bar, from which, attired in glittering tights, he would hang by one hand, one foot, or the back of his head, till 'the blood ceased to curdle in the veins of the awestruck crowd, and they gave vent to their feelings in cheer after cheer'. Then the master circus showman P. T. Barnum hired him for two seasons, and it was while Donaldson was in this employ that he conceived the truly American (and lovely!) idea of arranging the world's first ever airborne wedding.

2

The First Aerial Wedding
By Dr. M. L. Amick
QUOTED IN: Lighter-than-Air Flight
EDITED BY Lt. Col. C. V. Glines, USAF
Franklin Watts, New York, 1965

This ascension and wedding were to have occurred Saturday, October 17, 1874, and the monster balloon *P. T. Barnum* was fully inflated for the ascension, but all of a sudden the balloon collapsed, and the wedding was necessarily postponed until Monday, October 19. The rent having been repaired, the balloon, the *P. T. Barnum*, was again filled and made the ascension.

It was a glorious afternoon when the Hippodrome performance closed, at half past four. It was one of those days on the very eve of Indian summer when we in southern Ohio rejoice that we live there. There was a slight haze hanging over the city that softened but did not obscure the sunlight. Both moon and sun were visible; the latter sinking, round and lurid, to the west; the former rising

19

pale, crescent-shaped, from the eastern hills. A lovelier wedding hour never dawned on a happy bride, and a bonnier bride never welcomed it than Miss Mary Elizabeth Walsh, equestrienne of Barnum's Roman Hippodrome, who was to wed in mid-air her affianced, Mr. Charles M. Colton, also of the great show. The attendants were Miss Anna Rosetta Yates, the beautiful and daring equestrienne, and Mr W. C. Coup, Mr Barnum's popular business manager. The officiating minister was Rev Howard B. Jeffries, of the Church of Christ, a branch of the Swedenborgians of Pittsburg. The audience at the great show poured out to see the ascension. Lincoln Park and all the adjoining space were filled with a multitude which numbered full fifty thousand people, and which made up, probably, the largest wedding party on record. Mr. Donaldson was ready promptly, dressed to kill, and with bridal favors. The *P. T. Barnum* was full almost to bursting with the best of gas. Her basket was trimmed with flags and flowers. The ensigns of America and Ireland hung gracefully from her. All that is unsightly or forbidding about a balloon was hidden by the decorating care of loving hands. Rare bouquets hung from the ropes and baskets of exotics swung from the 'look out'. Mr. D. S. Thomas, the best of 'press agents', had general charge, and not even the minutest detail was neglected. Mr. P. T. Barnum and his party were of the selected company admitted within the ropes. A pathway was kept clear for the bridal procession, at the head of which marched the magnificent Hippodrome band playing Mendelssohn's 'Wedding March'.

The suite entered the basket by a stepladder, and were soon all ready to go. Then Mr. Donaldson found that the balloon would lift seven people, and Mr. Thomas was taken in, boutonniere and all. The bridal group was as picturesque in itself as any we have ever witnessed, even when all the fashionable world has filled famous churches to feast their eyes upon a long, stylish marriage train.

The girls were fair, good and exquisitely arrayed. The blond bride wore a delicate pearl-colored silk, with bias folds and heavy trimmings of fringe and puffing in the back. Her hat was the graceful 'brigand', of the same shade as the dress, with rakish white feather and 'pearl' bird, the exquisite toilette being completed by white gloves in dainty hands. The bridesmaid's toilette was equally elegant, on a basis of black silk, her hat and other appointments being the same as the bride's. Mrs. Frank Whittaker, under Mr. Barnum's carte blanche, had superintended the toilettes, and they were perfect. The richly decorated aerial ship rose as gently and as gracefully and as beautiful as a child's soap bubble, and ascended in a direct perpendicular line, while tens of thousands of throats shot up their plaudits. The balloon drifted slightly to the northward and westward, and in ten minutes a parachute thrown out told that the nuptial ceremony had been completed. It was sent fluttering downward at the precise moment that the 'Amen' of the shorter Swedenborgian ceremony was uttered, as was subsequently ascertained. The airship took a famous trip. She was over a mile high before the ceremony was completed, and directly over the Twenty-Fourth Ward. Then, striking a different current, it floated gracefully across the northward portion of the city, the sun being reflected brightly from its new coat of varnish, and typifying to those below the happiness of those who were 'drifting' above. Well may the bride forever remember, and quote:

No more, no more the worldly shore
Upbraids me with its loud uproar!
With dreamful eyes my spirit lies
Under the walls of Paradise!

The voyagers went to where they could look down on Eden Park; then with another change of current, floated westward and settled gently down on an open lot near the residence of John Shillito, Esq. Their midair ceremony had been solemnly performed by Mr. Jeffries, whose tones never trembled as he spoke.

The minister made the following remarks after the ceremony:
'Marriage is not an earthly but a heavenly institution, belonging to higher realms of life, and as such is it revered by the enlightened; the greater the enlightenment of any country or community the greater the respect it accords marriage; as an institution above those of the world, merely, it is, then, most fitting that its solemnization should be celebrated far above the earth.'

21

Wilbur and Orville Wright were the two sons of a minor bishop, and the owners of a bicycle shop in the American Mid-west. They were somewhat eccentric: as children they had sworn a pact with their sister never to marry, and though she did, they kept their word. They were perhaps an unlikely pair to succeed in being the first men to fly under power and in a machine under full control – except that they were superb mechanics (they didn't just sell bikes, they made them) who built their own wind tunnel and the gasoline engine for their plane. They spent years researching flying; journeying every year to the sand dunes of the Carolina coast for a working vacation where they would try to fly their creations. They took photographs at every step, and kept voluminous notes and diaries. The first excerpt here is a diary entry for the day they first actually flew – completely deadpan prose exactly like that for the other thousands of days on which they experimented. The Wrights later complained that no one took any notice of their colossal achievement, and few believed it: seeing how unexcited they seem to have been about it themselves, who can blame their contemporaries?

The second excerpt concerns Wilbur's demonstrations of this new-fangled flying in France in 1908, when he utterly astonished the French at how good at it he had become. The advice to Orville, then preparing a demonstration for the U.S. Army back home, has an unlikely poignancy: Orville crashed and was seriously injured and his army passenger was killed.

3

FROM: *The Papers of Wilbur and Orville Wright*

EDITED BY   Marvin W. McFarland
McGraw Hill, New York, 1953

Orville Wright's Diary, 1903

Thursday, December 17, 1903
When we got up a wind between 20 and 25 miles was blowing from the north. We got the machine out early and put out the signal for the men at the station. Before we were quite ready, John

T. Daniels, W. S. Dough, A. D. Etheridge, W. C. Brinkley of Manteo, and Johnny Moore of Nags Head arrived. After running the engine and propellers a few minutes to get them in working order, I got on the machine at 10:35 for the first trial. The wind, according to our anemometers at this time, was blowing a little over 20 miles (corrected) 27 miles according to the Government anemometer at Kitty Hawk. On slipping the rope, the machine started off increasing in speed to probably 7 or 8 miles. The machine lifted from the truck just as it was entering on the fourth rail. Mr. Daniels took a picture just as it left the tracks. I found the control of the front rudder quite difficult on account of its being balanced too near the center and thus had a tendency to turn itself when started so that the rudder * was turned too far on one side and then too far on the other. As a result, the machine would rise suddenly to about 10 ft. and then as suddenly, on turning the rudder, dart for the ground. A sudden dart when out about 100 feet from the end of the tracks ended the flight. Time about 12 seconds (not known exactly as watch was not promptly stopped). The lever for throwing off the engine was broken, and the skid under the rudder cracked. After repairs, at 20 min. after 11 o'clock Will made the second trial. The course was about like mine, up and down but a little longer over the ground though about the same in time. Dist. not measured but about 175 ft. Wind speed not quite so strong. With the aid of the station men present, we picked the machine up and carried it back to the starting ways. At about 20 minutes till 12 o'clock I made the third trial. When out about the same distance as Will's, I met with a strong gust from the left which raised the left wing and sidled the machine off to the right in a lively manner. I immediately turned the rudder to bring the machine down and then worked the end control. Much to our surprise, on reaching the ground the left wing struck first, showing the lateral control of this machine much more effective than on any of our former ones. At the time of its sidling it had raised to a height of probably 12 to 14 feet. At just 12 o'clock Will started on the fourth and last trip. The machine started off with its ups and downs as it had before, but by the time he had gone over

* What we would call the (front) elevator.

23

three or four hundred feet he had it under much better control, and was traveling on a fairly even course. It proceeded in this manner till it reached a small hummock out about 800 feet from the starting ways, when it began its pitching again and suddenly darted into the ground. The front rudder frame was badly broken up, but the main frame suffered none at all. The distance over the ground was 852 feet in 59 seconds. The engine turns was 1071, but this included several seconds while on the starting ways and probably about a half second after landing. The jar of landing had set the watch on machine back so that we have no exact record for the 1071 turns. Will took a picture of my third flight just before the gust struck the machine. The machine left the ways successfully at every trial, and the tail was never caught by the truck as we had feared.

After removing the front rudder, we carried the machine back to camp. We set the machine down a few feet west of the building, and while standing about discussing the last flight, a sudden gust of wind struck the machine and started to turn it over. All rushed to stop it. Will who was near one end ran to the front, but too late to do any good. Mr. Daniels and myself seized spars at the rear, but to no purpose. The machine gradually turned over on us. Mr. Daniels, having had no experience in handling a machine of this kind, hung on to it from the inside, and as a result was knocked down and turned over and over with it as it went. His escape was miraculous, as he was in with the engine and chains. The engine legs were all broken off, the chain guides badly bent, a number of uprights, and nearly all the rear ends of the ribs were broken. One spar only was broken.

After dinner we went to Kitty Hawk to send off telegram to M. W. While there we called on Capt. and Mrs. Hobbs, Dr. Cogswell and the station men.

Orville Wright to Bishop Milton Wright [*Telegram*]
Kitty Hawk, December 17, 1903
Success four flights Thursday morning all against twenty-one mile wind started from level with engine power alone average speed through air thirty-one miles longest 57 seconds inform press home Christmas.

Wilbur Wright to Orville

Le Mans, August 15, 1908

Last Saturday I took the machine out for the first time and made a couple of circles. On Monday I made two short flights. In the first I wound up with a complete $\frac{3}{4}$ of a circle with a diameter of only 31 yards, by measurement, *and landed with the wings level*. I had to turn suddenly as I was running into trees and was too high to land and too low to go over them. In the second flight I made an 'eight' and landed at the starting point. The newspapers and the French aviators nearly went wild with excitement. Bleriot & Delagrange were so excited they could scarcely speak, and Kapperer could only gasp, and could not talk at all. You would have almost died of laughter if you could have seen them. The French newspapers, *Matin, Journal, Figaro, L'Auto, Petit-Journal, Petit Parisien,* &c., give reports fully as favorable as the *Herald*. You never saw anything like the complete reversal of position that took place, after two or three little flights of less than two minutes each. Deutsch telegraphed to inquire whether he could have the 100,000 fr. stock and definitely took it. The English Mercedes-Daimler Co. have written to know whether they can have England on same terms as the published Weiler contract. They also would like to arrange the German business, I presume through the German Daimler Co. I have asked them to send a man to talk over matters.

We certainly cannot kick on the treatment the newspapers have given us, even *Les Sports* has acknowledged itself mistaken. I thought the first accident might bring out a different turn from some of them, but there has been little tendency that way yet. On Thursday I made a blunder in landing and broke three spars and all but one or two ribs in the left wings and three spars ends of the central section, and one skid runner. It was a pretty bad smash up, but Kapperer who was present pronounced it as fine a demonstration of the practicability of flying as the flights themselves. . . .

In your flights at Washington I think you should be careful to begin practice in calms and *keep well above the ground*. You will probably be unable to cut as short curves as I do here, but you will have it easier on your speed test in a straight line. . . .

Be awfully careful in beginning practice and go slowly.

25

Wilbur Wright to his father

Le Mans, August 15, 1908

In my experiments I have my two men and in addition a special corps of high priced assistants consisting of M. Bollée, & M. Pellier the richest men in Le Mans, who come out every day and work twice as hard as common laborers. Pellier is one of the large manufacturers of canned goods in France and has factories in a number of different towns. He has sent me for my lunches all kinds of the finest sardines, anchovies, asparagus, &c, &c. you ever saw. The people of Le Mans are exceedingly friendly and proud of the fame it is giving their town. I am in receipt of bouquets, baskets of fruit &c., almost without number. The men down at Bollée's shop have taken up a collection to buy me a testimonial of their appreciation. They say that I, too, am a workman. I wish Orville could have been here, but I presume he will find similar treatment at Washington. Only he will miss seeing Bleriot, Delagrange & Kapperer so excited that they could scarcely talk, gasping that nothing like our flights had ever been seen in France. Yet I had made flights of only a few minutes each so far.

Wilbur Wright to his sister

Le Mans, August 22, 1908

The newspapers have told every thing I have done and still more that I have not done so you know pretty well how things are going. The way the French have thrown up the sponge and made a grab for the band wagon is a great surprise. M. Peyrey the aeronautical editor of *Auto* has been favorable to us ever since I met him on my arrival in France, and he has almost outdone himself since I began flying. Several others are not much behind. All question as to who originated the flying machine has disappeared. The furor has been so great as to be troublesome. I cannot even take a bath without having a hundred or two people peeking at me. Fortunately every one seems to be filled with a spirit of friendliness and this makes it possible to deal with them without a fuss. . . .

We have even been set to music, and every one is singing a song 'Il Vol' [he flies] of which I will send you a copy as soon as I can get one. You really can have no comprehension of the enthusiasm

26

with which the flights have been greeted, especially in France, but almost equally in the rest of Europe. The dangerous feature is that they will be too enthusiastic and that a reaction will set in. I do not like such conditions. But in any event it has resulted in advertising us ten times more than anyone has been advertised before in connection with flying and settled all questions of priority.

Wilbur Wright to Orville

Le Mans, August 25, 1908

The excitement aroused by the short flights I have made is almost beyond comprehension. The French have simply become wild. Instead of doubting that we could do anything they are ready to believe that we can do everything. So the present situation is almost as troublesome as the former one. People have flocked here from all over Europe, and as I wish to practice rather than give exhibitions it is a little embarrassing. But I tell them plainly that I intend for the present to experiment only under the most favorable conditions. If the wind is more than five miles an hour I stay in. In a calm you can detect a mismovement instantly, but in winds you do not know at first whether the trouble is due to mistakes or to wind gusts. I advise you most earnestly to stick to calms till after you are sure of yourself. Don't go out even for all the officers of the government unless you would go equally if they were absent. *Do not let yourself be forced into doing anything before you are ready.* Be very cautious and proceed slowly in attempting flights in the middle of the day when wind gusts are frequent. Let it be understood that you wish to practice rather than give demonstrations and that you intend to do it in your own way. Do not let people talk to you all day and all night. It will wear you out, before you are ready for real business. Courtesy has limits. If necessary appoint some hour in the day time and refuse absolutely to receive visitors even for a minute at other times. Do not receive *any one* after 8 o'clock at night. . . .

A few days ago I was presented with a medal of the International Peace Society of which Baron d'Estournelles de Constant is president. Another for you was also given into my charge.

That English crowd, Daimler Mercedes, is ready to make a

contract similar to the Weiler contract but at a higher price. However I fear that they are more interested in selling stock than doing regular business and I am waiting to make further investigations, and consider other offers.

That visionary architect Le Corbusier published a book on aircraft in 1935. It is principally photographs, and very artistic ones, too, but there is a short introductory text written by the great man himself. He had been lucky enough to catch a glimpse of an exceedingly bold and romantic young Russian nobleman making the first ever aeroplane flight right into a city, through the turbulence high above the roofs and spires of Paris itself. Le Corbusier was thrilled.

4

FROM: *Aircraft*
BY Le Corbusier
The Studio, London, 1935

One night in the spring of 1909, from my student's garret on the Quai St. Michel I heard a noise which for the first time filled the entire sky of Paris. Until then men had been aware of one voice only from above – bellowing or thundering – the voice of the storm. I craned my neck out of the window to catch sight of this unknown messenger. The Comte de Lambert, having succeeded in 'taking off' at Juvisy, had descended towards Paris and circled the Eiffel Tower at a height of 300 metres. It was miraculous, it was mad! Our dreams then could turn into reality, however daring they might be.

It was a great joy, that night in Paris.

In spring, 1909, men had captured the chimera and driven it above the city.

You crashed a lot if you were one of aviation's pioneers, but you were seldom high enough to have far to fall, or going fast enough do yourself much mischief. Only nine people had been killed in aeroplane accidents by the end of 1909, but thirty-two died in 1910, and thereafter the accident record grew alarmingly. So you picked yourself up, collected your wreckage and went back to your shed to rebuild it – or begin a new design. Of course you studied most attentively any other machine you encountered, to see if it embodied any ideas worth pinching.

I suppose your enthusiasm for aviation fluctuated sharply, from minimal just after you'd had a crash, to wild and unbounded if you ever watched someone else actually flying.

The following account is by an English experimenter, describing the end of his first attempt to fly. The author was a bird-lover who first walked-out his proposed take-off path across the meadow to be sure it was free of larks' nests. He was Geoffrey de Havilland, who went on to fame and fortune as the designer of many First World War types, and of the Moth lightplanes between the wars. His company originated the Mosquito bomber of the Second World War; the world's first jet airliners, the tragic Comets; and the Tridents in which we fly around Europe today. He also endured personal tragedy, for two of his three sons were killed in aeroplane accidents. The de Havilland company no longer exists, having been absorbed into the Hawker Siddeley Group – now itself part of the British Aerospace Corporation.

5

FROM: *Sky Fever*

BY   Sir Geoffrey de Havilland
Hamish Hamilton, London, 1961

I held the handkerchief up to the wind and Hearle and I watched it anxiously. So long as it did not blow out more than four or five degrees from the vertical we knew it was all right. It hardly stirred in the light breeze. Flying conditions were ideal. We started up the four-cylinder engine and the fragile framework of our box kite shuddered because the balance of the engine was not too good. It seemed scarcely possible that this delicate assembly of timber, piano wire and doped fabric that I had

in a 'dog-fight', it was more a question of catch as catch can. A pilot would go down on the tail of a Hun, hoping to get him in the first burst; but he would not be wise to stay there, for another Hun would almost certainly be on *his* tail hoping to get him in the same way. Such fights were really a series of rushes, with momentary pauses to select the next opportunity – to catch the enemy at a disadvantage, or separated from his friends.

But apart from fighting, when twenty or thirty scouts were engaged, there was always a grave risk of collision. Machines would hurtle by, intent on their private battles, missing each other by feet. So such fighting demanded iron nerves, lightning reactions, snap decisions, a cool head, and eyes like a bluebottle, for it all took place at high speed and was three dimensional.

At this sort of sharpshooting some pilots excelled others; but in all air fighting (and indeed in every branch of aerial warfare) there is an essential in which it differs from the war on the ground: its absolute coldbloodedness. You cannot lose your temper with an aeroplane. You cannot 'see red', as a man in a bayonet fight. You certainly cannot resort to 'Dutch' courage. Any of these may fog your judgment – and that spells death.

Often at high altitudes we flew in air well below freezing point. Then the need to clear a jam or change a drum meant putting an arm out into an icy 100 m.p.h. wind. If you happened to have bad circulation (as I had), it left the hand numb, and since you could not stamp your feet, swing your arms, or indeed move at all, the numbness would spread to the other hand and sometimes to the feet as well. In this condition we often went into a scrap with the odds against us – they usually were against us, for it was our job to be 'offensive' and go over into enemy country looking for trouble – coldbloodedly in the literal sense; but none the less we had to summon every faculty of judgment and skill to down our man or, at the worst, to come out of it alive ourselves. So, like duelling, air fighting required a set steely courage, drained of all emotion, fined down to a tense and deadly effort of will. The Angel of Death is less callous, aloof and implacable than a fighting pilot when he dives.

There were, of course, emergency methods, such as standing the machine on its tail and holding it there just long enough to get one

35

good burst into the enemy above you; but nobody would fight that way if he could help it, though, actually, an SE5 pilot could do the same thing by pulling his top gun down the quadrant. He could then fire it vertically upward while still flying level.

This was how Beery Bowman once got away from an ugly situation. He had been scrapping a couple of Huns well over the other side of the lines. He managed to crash one of them, but in so doing exhausted the ammunition of his Vickers gun: his Lewis was jammed. The other Hun pursued him and forced him right down on to the 'carpet' – about a hundred feet from the ground. There was nothing to do but to beat it home. The Hun, out to avenge the death of his friend, and having the advantage of speed and height over Beery, chivvied him back to the lines, diving after him, bursting his gun, zooming straight up again, hanging there for a moment in a stall, and falling to dive again. He repeated this several times (he must have been a rotten shot) while Beery, with extraordinary coolness and presence of mind, pulled down his Lewis gun and managed to clear the jam. The next time the Hun zoomed, Beery throttled right down and pulled back to stalling speed. The result was that when the Hun fell out of his zoom, Beery was not ahead of him as before, but beneath him. As the Hun dropped into his dive Beery opened fire with his Lewis gun, raking the body above him with a long burst. The Hun turned over on his back, dived, and struck the ground, bursting into flames. Beery laconically continued his way home. He was awarded the D.S.O.

With the exception of Ball, most crack fighters did not get their Huns in dog-fights. They preferred safer means. They would spend hours synchronizing their guns and telescopic sights so that they could do accurate shooting at, say, two or three hundred yards. They would then set out on patrol, alone, spot their quarry (in such cases usually a two-seater doing reconnaissance or photography), and carefully manoeuvre for position, taking great pains to remain where they could not be seen, *i.e.* below and behind the tail of the enemy. From here, even if the Hun observer did spot them, he could not bring his gun to bear without the risk of shooting away his own tail plane or rudder. The stalker would not hurry after his quarry, but keep a wary eye to see he was not about to be attacked himself. He would gradually draw nearer, always in the blind spot, sight his guns very carefully, and then one long deadly burst would do the trick.

Such tactics as those were employed by Captain McCudden, V.C., D.S.O., amd also by the French ace, Guynemer. Both of them, of course, were superb if they got into a dog-fight; but it was in such fighting that they were both ultimately killed.

Typical logbook entries:

'5/5/17. Offensive patrol: twelve thousand feet. Hoidge, Melville and self on voluntary patrol. Bad Archie over Douai. Lost Melville in cloud and afterwards attacked five red scouts. Sheered off when seven others came to their assistance. Two against twelve "no bon". We climbed west and they east, afterwards attacked them again, being joined by five Tripehounds, making the odds seven to twelve. Think I did in one, and Hoidge also did in one. Both granted by Wing.'

'7/5/17. Ran into three scouts east of Cambrai. Brought one down. Meintjies dived, but his gun jammed, so I carried on and finished him. Next fired on two-seater this side lines, but could not climb up to him. Went up to Lens, saw a two-seater over Douai, dived and the others followed. Fixed him up. Afterwards this confirmed by an FE2d, who saw burst into flames. Tackled three two-seaters who beat it east and came home. Good day!'

The squadron was doing well in Huns. Ball came back every day with a bag of one or more. Besides his SE5 he had a Nieuport scout, the machine in which he had done so well the previous year. He had a roving commission and, with two machines, was four hours a day in the air. Of the great fighting pilots his tactics were the least cunning. Absolutely fearless, the odds made no difference to him. He would always attack, single out his man, and close. On several occasions he almost rammed the enemy, and often came back with his machine shot to pieces.

One morning, before the rest of us had gone out on patrol, we saw him coming in rather clumsily to land. He was not a stunt pilot, but flew very safely and accurately, so that, watching him, we could not understand his awkward floating landing. But when he taxied up to the sheds we saw his elevators were flapping loose – controls had been completely shot away! He had flown back from the lines and made his landing entirely by winding his adjustable tail up and down! It was incredible he had not crashed. His oil tank had been riddled, and his face and the whole nose of the machine were running with black castor oil. He was so angry at being shot up like this that he walked straight to the sheds, wiped the oil off his shoulders and face with a rag, ordered out his Nieuport, and within two hours was back with yet another Hun to his credit!

Ball was a quiet, simple little man. His one relaxation was the violin, and his favourite after-dinner amusement to light a red magnesium flare outside his hut and walk round it in his pyjamas, fiddling! He was meticulous in the care of his machines, guns, and in the examination of his ammunition. He never flew for amusement. The only trips he took, apart from offensive patrols, were the minimum requisite to test his engines or fire at the ground target sighting his guns. He never boasted or criticized, but his example was tremendous.

The squadron sets out eleven strong on the evening patrol. Eleven chocolate-coloured, lean, noisy bullets, lifting, swaying, turning, rising into formation – two fours and a three – circling and climbing away steadily towards the lines. They are off to deal with Richthofen and his circus of Red Albatrosses.

The May evening is heaving with threatening masses of cumulus cloud, majestic skyscapes, solid-looking as snow mountains, fraught with caves and valleys, rifts and ravines – strange and secret pathways in the chartless continents of the sky. Below, the land becomes an ordnance map, dim green and yellow, and across it go the Lines, drawn anyhow, as a child might scrawl with a double pencil. The grim dividing Lines! From the air robbed of all significance.

Steadily, the body of scouts rises higher and higher, threading its way between the cloud precipices. Sometimes, below, the streets of a village, the corner of a wood, a few dark figures moving, glide into view like a slide into a lantern and then are hidden again.

But the fighting pilot's eyes are not on the ground, but roving endlessly through the lower and higher reaches of the sky, peering anxiously through fur-goggles to spot those black slow-moving specks against land or cloud which mean full throttle, tense muscles, held breath, and the headlong plunge with screaming wires – a Hun in the sights, and the tracers flashing.

A red light curls up from the leader's cockpit and falls away. Action! He alters direction slightly, and the patrol, shifting throttle and rudder, keep close like a pack of hounds on the scent. He has seen, and they see soon, six scouts three thousand feet below. Black crosses! It seems interminable till the eleven come within diving distance. The pilots nurse their engines, hard-minded and set, test their guns and watch their indicators. At last the leader sways sideways, as a signal that each should take his man, and suddenly drops.

Machines fall scattering, the earth races up, the enemy patrol, startled, wheels and breaks. Each his man! The chocolate thunderbolts take sights, steady their screaming planes, and fire. A burst, fifty rounds – it is over. They have overshot, and the enemy, hit or missed, is lost for the moment. The pilot steadies his stampeding mount, pulls her out with a firm hand, twisting his head right and left, trying to follow his man, to sight another, to back up a friend in danger, to note another in flames.

But the squadron plunging into action had not seen, far off, approaching from the east, the rescue flight of Red Albatrosses patrolling above the body of machines on which they had dived, to guard their tails and second them in the battle. These, seeing the maze of wheeling machines, plunge down to join them. The British scouts, engaging and disengaging like flies circling at midday in a summer room, soon find the newcomers upon them. Then, as if attracted by some mysterious power, as vultures will draw to a corpse in the desert, other bodies of machines swoop down from the peaks of the cloud mountains. More enemy scouts, and, by good fortune, a flight of Naval Triplanes.

But, nevertheless, the enemy, double in number, greater in power and fighting with skill and courage, gradually overpower the British, whose machines scatter, driven down beneath the scarlet German fighters.

It would be impossible to describe the action of such a battle. A pilot, in the second between his own engagements, might see a Hun diving vertically, an SE5 on his tail, on the tail of the SE another Hun, and above him again another British scout. These four, plunging headlong at two hundred miles an hour, guns crackling, tracers streaming, suddenly break up. The lowest Hun plunges flaming to his death, if death has not taken him already. His victor seems to stagger, suddenly pulls out in a great leap, as a trout leaps on the end of a line, and then, turning over on his belly, swoops and spins in a dizzy falling spiral with the earth to end it. The third German zooms veering, and the last of that meteoric quartet follows bursting. . . . But such a glimpse, lasting perhaps ten seconds, is broken by the sharp rattle of another attack. Two machines approach head-on at breakneck speed, firing at each other, tracers whistling through each other's planes, each slipping sideways on his rudder to trick the other's gun fire. Who will hold longest? Two hundred yards, a hundred, fifty, and then, neither hit, with one accord they fling their machines sideways, bank and circle, each striving to bring his gun on to the other's tail, each glaring through goggle eyes, calculating, straining, wheeling, grim, bent only on death or dying.

But, from above, this strange tormented circling is seen by another Hun. He drops. His gun speaks. The British machine, distracted by the sudden unseen enemy, pulls up, takes a burst through the engine, tank and body, and falls bottom uppermost down through the clouds and the deep unending desolation of the twilight sky.

The game of noughts and crosses, starting at fifteen thousand feet above the clouds, drops in altitude engagement by engagement. Friends and foes are scattered. A last SE, pressed by two Huns, plunges and wheels, gun-jammed, like a snipe over marshes, darts lower, finds refuge in the ground mist, and disappears.

Now lowering clouds darken the evening. Below, flashes of gun fire stab the veil of the gathering dusk. The fight is over! The battlefield shows no sign. In the pellucid sky, serene cloud mountains mass and move unceasingly. Here where guns rattled and death plucked the spirits of the valiant, this thing is now as if it had never been! The sky is busy with night, passive, superb, unheeding.

Of the eleven scouts that went out that evening, the 7th of May, only five of us returned to the aerodrome.

The mess was very quiet that night. The Adjutant remained in his office, hoping against hope to have news of the six missing pilots and, later, news did come through that two had been forced down, shot in the engine, and that two others had been wounded.

But Ball never returned. I believe I was the last to see him in his red-nosed SE going east at eight thousand feet. He flew straight into the white face of an enormous cloud. I followed. But when I came out on the other side, he was nowhere to be seen. All next day a feeling of depression hung over the squadron. We mooned about the sheds, still hoping for news. The day after that hope was given up. I flew his Nieuport back to the Aircraft Depot.

It was decided to go over to Douai and drop message-bags containing requests, written in German, for news of his fate. We crossed the lines at thirteen thousand feet. Douai was renowned for its anti-aircraft. They were not to know the squadron was in mourning, and made it hot for us. The flying splinters ripped the planes. Over the town the message-bags were dropped, and the formation returned without encountering a single enemy machine.

Later, word was received that Ball had been killed. He was posthumously awarded the Victoria Cross.

Not many people know that the Bloody Red Baron, Manfred von Richthofen, wrote and published an autobiography. Introducing its 1918 English translation, that crusty old aviation journalist C. G. Grey, editor of The Aeroplane, *said the work 'gives one the general impression of the writings of a gentleman, prepared for publication by a hack journalist,' and added that it had obviously been 'carefully censored' and 'obviously touched up here and there for propagandist purposes'. I've always felt that this book is very much von Richthofen himself talking. It gives a powerful impression of the man, and not a nice one: rather, a cold-blooded killer devoid of imagination and possessed of a particularly nasty sense of humour.*

I have never been able to admire the real, as opposed to the legendary, Red Baron, except for his bravery and marksmanship. At his death, he had shot down eighty Allied aircraft – easily the highest score in the whole war – and become Germany's foremost war hero. The Allies (on whose side of the line he ultimately fell) gave him a funeral with full military honours; but no one knows to this day who actually got him – the Canadian pilot Captain Roy Brown who was on his tail in a Sopwith Camel, or an Australian machine gunner on the ground. Towards the end Richthofen seemed to grow contemptuous of danger, so his death was surely inevitable.

In the following excerpt, von Richthofen comments widely on the relative abilities of the British and French aviators, and describes with appalling arrogance how he was himself once shot down (unhurt) and was then rescued by a particularly boorish German army officer, in whose subsequent discomforture von Richthofen clearly rejoiced.

7

FROM: *The Red Air Fighter*

BY   Manfred Freiherr von Richthofen
     Aeroplane & General, London, 1918

It occurred to me to have my packing case painted all over in staring red. The result was that everyone got to know my red bird. My opponents also seemed to have heard of the colour transformation.

During a fight on quite a different section of the front I had the good fortune to shoot a Vickers' two-seater which was peacefully photographing the German artillery position. My friend the photographer had not the time to defend himself. He had to make haste to get down upon firm ground, for his machine began to give suspicious indications of fire. When we notice that phenomenon, we say: 'He stinks!' As it turned out, it was really so. When the machine was coming to earth it burst into flames.

I felt some human pity for my opponent and had resolved not to cause him to fall down but merely to compel him to land. I did so particularly because I had the impression that my opponent was wounded, for he did not fire a single shot.

When I had got down to an altitude of about 1,500 feet engine trouble compelled me to land without making any curves. The result was very comical. My enemy with his burning machine landed smoothly, while I, his conqueror, came down next to him in the barbed wire of our trenches and my machine overturned.

The two Englishmen who were not a little surprised at my collapse, greeted me like sportsmen. As mentioned before, they had not fired a shot, and they could not understand why I had landed so clumsily. They were the first two Englishmen whom I had brought down alive. Consequently, it gave me particular pleasure to talk to them. I asked them whether they had previously seen my machine in the air, and one of them replied, 'Oh, yes. I know your machine very well. We call it "Le Petit Rouge." '

*   *   *

I was trying to compete with Boelcke's Squadron. Every evening we compared our bags. However, Boelcke's pupils are smart rascals. I cannot get ahead of them. The utmost one can do is to draw level with them. The Boelcke *stäffel* has an advantage over my people of 100 aeroplanes downed. I must allow them to retain it. Everything depends on whether we have for opponents those French tricksters or those daring fellows the English. I prefer the English. Frequently the daring of the latter can only be described as stupidity. In their eyes it may be pluck and bravery.

The great thing in air fighting is that the decisive factor does not lie in trick flying but solely in the personal ability and energy of the aviator. A flying man may be able to loop and do all the tricks imaginable and yet he may not succeed in shooting down a single enemy. In my opinion the aggressive spirit is everything, and that spirit is very strong in us Germans. Hence we shall always retain the domination of the air.

The French have a different character. They like to set traps and to attack their opponents unawares. That cannot easily be done in the air. Only a beginner can be caught, and one cannot set traps, because an aeroplane cannot hide itself. The invisible aeroplane has not yet been discovered. Sometimes, however, the Gallic blood asserts itself. Then Frenchmen will then attack. But the French attacking spirit is like bottled lemonade. It lacks tenacity.

In Englishmen, on the other hand, one notices that they are of Germanic blood. Sportsmen easily take to flying, but Englishmen see in flying nothing but a sport. They take a perfect delight in looping the loop, flying on their back, and indulging in other tricks for the benefit of our soldiers in the trenches. All these tricks may impress people who attend a Sports Meeting, but the public at the battle-front is not as appreciative of these things. It demands higher qualifications than trick flying. Therefore, the blood of English pilots will have to flow in streams.

*   *   *

I have had an experience which might perhaps be described as being shot down. At the same time, I call it being shot down only when one falls down. Today I got into trouble, but I escaped with a whole skin.

I was flying with the squadron and noticed an opponent who also was flying in a squadron. It happened above the German artillery position in the neighbourhood of Lens. I had to fly quite a distance to get there. It tickles one's nerves to fly towards the enemy, especially when one can see him from a long distance and when several minutes must elapse before one can start fighting. I imagine that at such a moment my face turns a little pale, but unfortunately I have never had a mirror with me. I like that feeling, for it is a wonderful nerve stimulant.

One observes the enemy from afar. One has recognised that his squadron is really an enemy formation. One counts the number of the hostile machines and considers whether the conditions are favourable or unfavourable. A factor of enormous importance is whether the wind forces one away from our front or towards our front. For instance, I once shot down an Englishman. I fired the fatal shot above the English position. However, the wind was so strong that his machine came down close to the German kite-balloons.

We Germans had five machines. Our opponents were three times as numerous. The English flew about like midges. It is not easy to disperse a swarm of machines which fly together in good order. It is impossible for a single machine. It is extremely difficult for several aeroplanes, particularly if the difference in number is as great as it was in this case. However, one feels such a superiority over the enemy that one does not doubt for a moment of success.

The aggressive spirit, the offensive, is the chief thing everywhere in war, and the air is no exception. * However, the enemy had the same idea. I noticed that at once. As soon as they noticed us they turned round and attacked us. Now we five had to look sharp. If one of them should fall there might be a lot of trouble for all of us. We went closer together and allowed the foreign gentlemen to approach us.

I watched whether one of the fellows would hurriedly take leave of his colleagues. One of them was stupid enough to depart alone. I could reach him and I said to myself, 'That man is lost!' Shouting aloud, I went after him. I came up to him, or at least was getting very near him. He started shooting prematurely, which showed that he was nervous. So I said to myself, 'Go on shooting. You won't hit me.' He shot with a kind of munition which ignites. So I could see his shots passing me. I felt as if I were sitting in front of a gigantic watering pot. The sensation was not pleasant. Still, the English usually shoot with this beastly stuff, and so we must try and get accustomed to it. One can get accustomed to anything. At the moment, I think, I laughed aloud. But soon I got a lesson. When I had approached the Englishman quite closely, when I had come to a distance of about 300 feet, I got ready for firing, aimed and gave a few trial shots. The machine-guns were in order. The decision would be there before long. In my mind's eye I saw my enemy dropping.

* The doctrine of Clausewitz and of the leading German military writers.

My former excitement was gone. In such a position one thinks quite calmly and collectedly and weighs the probabilities of hitting and of being hit. Altogether the fight itself is the least exciting part of the business as a rule. He who gets excited in fighting is sure to make mistakes. He will never get his enemy down. Besides calmness is, after all, a matter of habit. At any rate, in this case I did not make a mistake. I approached my man up to within fifty yards. I fired some well-aimed shots and thought that I was bound to be successful. That was my idea. But suddenly I heard a tremendous bang when I had scarcely fired ten cartridges, and presently again something hit my machine. It became clear to me that I had been hit, or rather my machine. At the same time I noticed a fearful stench of petrol, and I observed that the motor was running slack. The Englishman noticed it too, for he started shooting with redoubled energy, while I had to stop it.

I went right down. Instinctively I switched off the engine, and indeed it was high time to do this. When one's petrol tank has been holed and when the infernal liquid is squirting around one's legs the danger of fire is very great. One has in front an explosion engine of more than 150 h.p. which is red hot. If a single drop of petrol should fall on it the whole machine would be in flames.

I left in the air a thin white cloud. I knew its meaning from my enemies. Its appearance is the first sign of a coming explosion. I was at an altitude of 9,000 feet, and had to travel a long distance to get down. By the kindness of Providence my engine stopped running. I have no idea with what rapidity I went downward. At any rate the speed was so great that I could not put my head out of the machine without being pressed back by the rush of air.

Soon I had lost sight of my enemy. I had only time to see what my four comrades were doing while I was dropping to the ground. They were still fighting. Their machine-guns and those of their opponents could be heard. Then I notice a rocket. Is it a signal of the enemy? No, it cannot be. The light is too great for a rocket. Evidently a machine is on fire. What machine? The burning machine looks exactly as if it were one of our own. No! Praise the Lord, it is one of the enemy's! Who can have shot him down? Immediately afterwards a second machine drops out and falls perpendicularly to the ground, turning, turning, turning exactly as I did, but suddenly it recovers its balance. It flies straight towards me. It also is an Albatros. No doubt it had the same experience as I had.

I had fallen to an altitude of perhaps 1,000 feet, and had to look out for a landing. Now, such a sudden landing usually leads to breakages, and these are occasionally serious. I found a meadow. It was not very large, but it just sufficed if I used due caution. Besides it was favourably situated on the high road near Hénin-Liétard. There I meant to land.

Everything went as desired, and my first thought was, 'What has become of the other fellow?' He landed a few kilometres from the spot where I had come to the ground.

I had ample time to inspect the damage. My machine had been hit a number of times. The shot which caused me to give up the fight had gone through both the petrol tanks. I had not a drop of petrol left, and the engine itself also had been damaged by shots. It was a pity, for it had worked so well.

I let my legs dangle out of the machine, and probably made a very silly face. In a moment I was surrounded by a large crowd of soldiers. Then came an officer. He was quite out of breath. He was terribly excited. No doubt something fearful had happened to him. He rushed towards me, gasped for air and asked: 'I hope that nothing has happened to you? I have followed the whole affair, and am terribly excited! Good Lord, it looked awful!' I assured him that I felt quite well, jumped down from the side of my machine and introduced myself to him. Of course he did not understand a particle of my name. However, he invited me to go in his motor car to Hénin-Liétard, where he was quartered. He was an Engineer Officer.

We were sitting in the motor and were commencing our ride. My host was still extraordinarily excited. He jumped up, and asked: 'Good Lord, but where is your chauffeur?' At first I did not quite understand what he meant. Probably I looked puzzled. Then it dawned upon me that he thought that I was the observer of a two-seater, and that he asked after the fate of my pilot. I pulled myself together, and said in the driest tones: 'I always drive myself.' Of course the word 'drive' is absolutely taboo among the flying men.

An aviator does not drive, he flies. In the eyes of the kind gentleman I had obviously lost caste when he discovered that I

47

'drove' my own aeroplane. The conversation began to slacken.

We arrived at his quarters. I was still dressed in my dirty and oily leather jacket, and had round my neck a thick wrap. On our journey he had of course asked me a tremendous number of questions. Altogether he was far more excited than I was.

When we got to his diggings he forced me to lie down on the sofa, or at least he tried to force me because, he argued, I was bound to be terribly done up through my fight. I assured him that this was not my first aerial battle, but he did not, apparently, give me much credence. Probably I did not look very martial.

After we had been talking for some time he asked me of course the celebrated question: 'Have you ever brought down a machine?' As I said before he had probably not understood my name. So I answered nonchalantly, 'Oh, yes. I have done so now and then.' He replied: 'Indeed! Perhaps you have shot down two?' I answered: 'No. Not two, but twenty-four.' He smiled, repeated his question and gave me to understand that, when he was speaking about shooting down an aeroplane, he meant not shooting *at* an aeroplane, but shooting *into* an aeroplane in such a manner that it would fall to the ground and remain there. I immediately assured him that I entirely shared his conception of the meaning of the words 'shooting down'.

Now I had completely lost caste with him. He was convinced that I was a fearful liar. He left me sitting where I was, and told me that a meal would be served in an hour. If I liked I could join in. I accepted his invitation, and slept soundly for an hour. Then we went to the Officers' Club. Arrived at the Club I was glad to find that I was wearing the Ordre pour le Mérite.

Unfortunately I had no uniform jacket underneath my greasy leather coat, but only a waistcoat. I apologised for being so badly dressed. Suddenly my good officer discovered on me the Ordre pour le Mérite. He was speechless with surprise, and assured me that he did not know my name. I gave him my name once more. Now it seemed to dawn upon him that he had heard my name before. He feasted me with oysters and champagne, and I did gloriously until at last my orderly arrived and fetched me with my car.

I learned from him before leaving his quarters that comrade Lübbert had once more justified his nick-name. He was generally called 'the bullet-catcher', for his machine suffered badly in every fight. Once it was hit sixty-four times. Yet he had not been wounded. This time he had received a glancing shot on the chest, and he was by this time in hospital. I flew his machine to port. Unfortunately this excellent officer, who promised to become another Boelcke, died a few weeks later a hero's death for the Fatherland.

In the evening I could assure my kind host of Hénin-Liétard that I had increased my 'bag' to twenty-five.

Aerial warfare with aeroplanes had actually begun in skirmishes before 1914, in Mexico, Morocco and the Balkans; the principal combatants in the Great War all started with tiny air arms in 1914 which, by 1918, had grown to huge organisations. There was rather less glamour in the air fighting than is popularly supposed; most pilots simply were dead within a few weeks of being sent to the front. And while many pilots would have told you that at least flying got you out of the mud and blood and gas and the cold damp misery of the trenches, most Tommies would have told you that nothing would induce them up in one of those machines. They saw too many of them come down, often in flames with the parachuteless pilot and observer being fried alive on the way.

There was precious little of the chivalry so often described in First World War fiction in real life. This excerpt from the memoirs of the German ace Ernst Udet, who flew with von Richthofen, and survived the war to become an aerobatic pilot of fame, a Hollywood stunt pilot, and later a minor minion of Hitler's entourage, describes one such unlikely incident. Film enthusiasts will remember that this little story was faithfully re-enacted in The Great Waldo Pepper.

8

FROM: *Ace of the Iron Cross*

BY Ernst Udet, translated by Richard K. Riehn
    reprinted Doubleday, New York, 1970

Jasta 15, which grew out of the old Single-Seater Combat Command Habsheim, has now only four aircraft, three sergeants, and myself as their leader. Almost always we fly alone. Only in this way can we fulfill our assigned duties.

Much is happening on the front. It is said the other side is preparing an offensive. The balloons are up every day, hanging in long rows in the summer sky, like a garland of fat-bellied clouds. It would be good if one of them were to burst. It would be a good warning to the others in addition, on just plain general principles.

I start early in the morning, so that I can have the sun at my back to stab down at the balloon. I fly higher than ever before. The altimeter shows five thousand meters. The air is thin and icy. The world below me looks like a gigantic aquarium. Above

Lierval, where Reinhold fell, an enemy pusher type is cruising around. Like a tiny water flea, he shovels his way through the air.

From the west, a small dot approaches fast. At first, small and black, it grows quickly as it approaches. A Spad, an enemy fighter. A loner like me, up here, looking for prey. I settle myself into my seat. There's going to be a fight.

At the same height, we go for each other, passing at a hair's breadth. We bank into a left turn. The other's aircraft shines light brown in the sun. Then begins the circling. From below, it might appear as though two large birds of prey were courting one another. But up here it's a game of death. He who gets the enemy at his back first is lost, because the single-seater with his fixed machine guns can only shoot straight ahead. His back is defenseless.

Sometimes we pass so closely I can clearly recognize a narrow, pale face under the leather helmet. On the fuselage, between the wings, there is a word in black letters. As he passes me for the fifth time, so close that his propwash shakes me back and forth, I can make it out: *'Vieux'* it says there – *vieux* – the old one. That's Guynemer's sign. *

Yes, only one man flies like this on our front, Guynemer, who has brought down thirty Germans, Guynemer, who always hunts alone, like all dangerous predators, who swoops out of the sun, downs his opponents in seconds, and disappears. Thus he got Puz away from me. I know it will be a fight where life and death hang in the balance.

I do a half loop in order to come down on him from above. He understands at once and also starts a loop. I try a turn, and Guynemer follows me. Once out of the turn, he can get me into his sights for a moment. Metallic hail rattles through my right wing plane and rings out as it strikes the struts.

I try anything I can, tightest banks, turns, side slips, but with lightning speed he anticipates all my moves and reacts at once. Slowly I realise his superiority. His aircraft is better, he can do more than I, but I continue to fight. Another curve. For a moment he comes into my sights. I push the button on the stick . . . the machine gun remains silent . . . stoppage!

* The entire inscription on Guynemer's Spad read *'Vieux Charles'* – 'Old Charlie'.

With my left hand clutched around the stick, my right attempts to pull a round through. No use – the stoppage can't be cleared. For a moment I think of diving away. But with such an opponent this would be useless. He would be on my neck at once and shoot me up.

We continue to twist and turn. Beautiful flying if the stakes weren't so high. I never had such a tactically agile opponent. For seconds, I forget that the man across from me is Guynemer, my enemy. It seems as though I were sparring with an older comrade over our own airfield. But this illusion lasts only for seconds.

For eight minutes we circle around each other. The longest eight minutes of my life. Now, lying on his back, he races over me. For a moment I have let go of the stick and hammer the receiver with both fists. A primitive expedient, but it helps sometimes.

Guynemer has observed this from above, he must have seen it, and now he knows what gives with me. He knows I'm helpless prey.

Again he skims over me, almost on his back. Then it happens: he sticks out his hand and waves to me, waves lightly, and dives to the west in the direction of his lines.

I fly home. I'm numb.

There are people who claim Guynemer had a stoppage himself then. Others claim he feared I might ram him in desperation. But I don't believe any of them. I still believe to this day that a bit of chivalry from the past has continued to survive. For this reason I lay this belated wreath on Guynemer's unknown grave.

*The more convivial pilots of the RFC tended to live each day as
though it were their last, with colossal parties and drunken binges to
fill the long evenings. In those before-TV, pre-radio days people still
made their own entertainment, and any young man of wit could strum
a piano and make up a simple ditty. Here are a couple of Songs They
Sang in the RFC, washing the words down with copious drafts of
wine and beer, and getting ever louder and more raucous towards the
last verse.*

9

The Dying Aviator

A young aviator lay dying,
  At the end of a bright summer's day (summer's day)
His comrades had gathered around him,
  To carry his fragments away.

The aeroplane was piled on his wishbone,
  His Lewis was wrapped round his head (his head).
He wore a spark plug in each elbow,
  'Twas plain he would shortly be dead.

He spat out a valve and a gasket,
  As he stirred in the sump where he lay (he lay).
And then to his wondering comrades
  These brave parting words did he say:

'Take the manifold out of my larynx,
  And the butterfly-valve off my neck (my neck).
Remove from my kidneys the camrods,
  There's a lot of good parts in this wreck.

'Take the piston rings out of my stomach,
  And the cylinders out of my brain (my brain).
Extract from my liver the crankshaft,
  And assemble the engine again.

'Pull the longeron out of my backbone,
  The turnbuckle out of my ear (my ear).
From the small of my back take the rudder –
  There's all of your aeroplane here.

'I'll be riding a cloud in the morning,
    With no rotary before me to cuss (to cuss).
Take the lead from your feet and get busy,
    There's another lad needing the bus.'

In Other Words

I was fighting a Hun in the heyday of youth,
    Or perhaps 'twas a Nieuport or Spad.
I put in a burst at a moderate range
    And it didn't seem any too bad.
For he put down his nose in a curious way,
And as I watched him I am happy to say:

CHORUS: He descended with unparalleled rapidity,
                His velocity 't would beat me to compute.
            I speak with unimpeachable veracity,
                With evidence complete and absolute.
            He suffered from spontaneous combustion
                As towards terrestrial sanctuary he dashed,
            And underwent complete disintegration,
                In other words – he crashed!

I was telling the tale when a message came through
    To say 'twas a poor R.E.8
The news somewhat dashed me, I rather supposed
    I was in for a bit of a hate.
The C.O. approached me, I felt rather weak,
For his face went all mottled, and when he did speak:

CHORUS: He straffed me with unmitigated violence,
                With wholly reprehensible abuse.
            His language in its blasphemous simplicity
                Was rather more exotic than abstruse,
            He mentioned that the height of his ambition
                Was to see your humble servant duly hung.
            I returned to the Home Establishment next morning,
                In other words – I was stung.

As a pilot in France I flew over the lines
  And there met an Albatros Scout.
It seemed that he saw me, or so I presumed;
  His manoeuvre left small room for doubt.
For he sat on my tail without further delay
Of my subsequent actions I think I may say:

CHORUS: My turns approximated to the vertical,
        I deemed it most judicious to proceed.
      I frequently gyrated on my axis
        And attained colossal atmospheric speed,
      I descended with unparalleled momentum,
        My propeller's point of rupture I surpassed,
      And performed the most astounding evolutions,
        In other words, —

I was testing a Camel on last Friday week,
  For the purpose of passing her out.
And before fifteen seconds of flight had elapsed
  I was filled with a horrible doubt,
As to whether intact I should land from my flight.
I half thought I'd crash – and I half thought quite right.

CHORUS: The machine it seemed to lack co-agulation,
        The struts and sockets didn't rendezvous,
      The wings had lost their super-imposition,
        Their stagger and their incidental, too!
      The fuselage developed undulations,
        The circumjacent fabric came unstitched,
      Instanter was reduction to components,
        In other words – she's pitched!

Anthony Fokker was a Dutchman, citizen of a country that was neutral in the First World War. He was also one of the pioneers of flying, designing and building his first aeroplane in Germany in 1910. He had several hairbreadth escapes as an exhibition pilot in those pre-war years. He stayed in Germany to design aircraft for the German military; his famous 1915 Eindecker, with its interrupter gear allowing the machine gun to fire through the propeller's arc, was the world's first real fighter aircraft. Despite 'constant struggles with the High Command, and continual battles with rival German airplane manufacturers, resentful because a Dutchman was designing planes that their factories had to build on a royalty basis . . . hedged about by intriguing competitors plotting my downfall by open or devious schemes . . .' Fokker produced many of the finest fighters of the war, making a great fortune in the process before he was twenty-eight, and finally succeeding in smuggling his wealth back to Holland.

Perhaps understandably he became something of a cynic towards life, war and the business of armaments manufacture; but he never lost his admiration for the pilots who flew his creations in battle. Here he describes the characters of the leading German aces.

10

FROM: *Flying Dutchman*

BY Anthony H. G. Fokker and Bruce Gould

Holt, Rinehart & Winston, New York, 1931

The contempt of the German flyers for death was only equalled by their love of life while they still had that precious possession. So complete was their disregard for the hazard of aerial combat, I sometimes thought they were hardly aware of its terrible dangers. Yet that could not be possible, for on every day they went hunting in the skies some members of the *Jagdstaffel* failed to return. When I met them in their headquarters at the Front they jested and sported as though the angel of death were not the permanent leader of their circus, and when they came to Berlin for a fortnight's holiday, they lived as riotously as though they hadn't a care in the world. That is, with a few exceptions, among them Richthofen. He was calm, cold, ambitious; a born leader of men and Germany's greatest ace.

Richthofen, Boelcke, and Immelmann, Germany's trio of aces, I knew intimately; as intimately at least as one knows men who, having stared at death so often, have learned to wear a mask lest an occasional human weakness betray their almost hypnotic gallantry. They were as different as men of the same breed can be. One by one I saw them die as I knew they must die, for they were in a contest not with a human opponent but with Time, the cruelest foe in the world. Judging their bravery by my own, I reckoned them supreme. Knowing the accuracy of the machine gun and the airplane in the hands of a skilled pilot, calculating the remote chance of surviving any prolonged campaign in the air, I would never have had the courage to face the enemy. Every man who went aloft was marked for death, sooner or later, once his wheels left the ground.

Max Immelmann, with Boelcke, was the first German pilot to win the Pour le Mérite, the Empire's highest decoration for military bravery. This medal originated by Frederick the Great, was colloquially called 'the blue Max', from its colour and Frederick's name. Its French title was due to the fact that the founder of the German Empire would only speak French. Immelmann was a serious, modest youngster, intensely interested in the technical details of flying. He was popular, and originally better known than Boelcke. He came to Berlin after his fourth or fifth victory and I took him to Schwerin for a tour of my factory. We talked little of abstract matters, but always of machine guns – he was an excellent shot – of aerial maneuvers, of the relative merits of one pursuit plane over another. He had eyes like a bird of prey, and a short, athletic body capable of standing the bombardment of nerves from which every flyer suffers when alone with his imagination. At no time did he drop a hint that he considered air fighting dangerous. As far as I might have known, he had not the slightest care in the world. He gained fifteen victories before he was killed June 18, 1916.

Almost as much mystery surrounds the manner of Immelmann's death as Guynemer's, which was never adequately explained. Immelmann's plane suddenly fell to the ground as he was flying near the German front lines. It was first given out that his Fokker fighter had failed in midair. This explanation naturally did not satisfy me, and I insisted on examining the remains of the wreck, and establishing the facts of his death. What I saw convinced me and others that the fuselage had been shot in two by shrapnel fire. The control wires were cut as by shrapnel, the severed ends bent in, not stretched as they would have been in an ordinary crash. The tail of the fuselage was found a considerable distance from the plane itself. As he was flying over the German lines there was a strong opinion in the air force that his comparatively still unknown monoplane type – which somewhat resembled a Morane-Saulnier – had been mistaken for a French plane. I was finally able to convince air headquarters sufficiently so that, while it was not stated that he had been shot down by German artillery – which would have horrified his millions of admirers – neither was the disaster blamed on the weakness of his Fokker plane. The air corps exonerated the Fokker plane unofficially, although as far as the public was concerned the whole episode was hushed up. Because of this investigation, however, silhouettes of all German types were sent to all artillery commanders to prevent a repetition of the Immelmann catastrophe.

Boelcke, the son of a Saxon schoolmaster, was of quite a different type, although like Immelmann intensely interested in the technical details of flying and aerial combat. In a desperate effort to save him from inevitable death, the High Command restricted his flying after his sixteenth victory in 1916, and sent him to Austria, Bulgaria, and Turkey to instruct others in airmanship. But he became so wearied of the relentless adulation showered on him that he begged leave to return to the Front. Until Lieutenant Boehme, of his *staffel*, collided with his plane in midair causing his wing to drop off, his victories mounted, reaching a total of forty before he died. Lieutenant Boehme, who was barely restrained from suicide in his grief, was later shot down in a dogfight.

Choosing the flying corps because an asthmatic affliction kept him from harder labour, Boelcke left the signal corps shortly before the War to enter the Halberstadt flying school. After seven weeks' training he became a pilot and the first of September, 1914, saw him flying over the Western Front as an observer. It was in June of 1915 that he obtained his first Fokker single-seater in company with Immelmann and began his career as an ace. Boelcke had charm, and a kindness of heart which extended itself even to the enemies he brought down. He spent much of his leisure motoring to hospitals to cheer up his wounded opponents, leaving some gift of cigarettes or other trifle as he departed. Richthofen, who worshipped Boelcke and learned many of his flying tricks from him, records the fact that 'it is a strange thing that everybody who met Boelcke imagined that he alone was his true friend. I have met about forty men, each of whom imagined that he alone had Boelcke's affection. Men whose names were unknown to Boelcke believed that he was particularly fond of them. Boelcke had not a personal enemy.' Yet no one had a better record of bravery. He died on October 28, 1916.

Richthofen, with whom I became very friendly, was an entirely different sort of flyer from the other two. Without the subconscious art which Boelcke and Immelmann possessed, he was slow to learn to fly, crashing on his first solo flight and only mastering the plane at last by sheer force of superior will. Time and again he escaped death by a miracle before he managed to conquer the unruly plane which later became his willing slave. A Prussian, son of a Junker family, Richthofen was imbued with the usual ideas of a young nobleman. He flew spectacularly in his series of all-red planes which became famous over the Western Front. Flaunting himself in the face of his enemies, he built up a reputation which perhaps somewhat daunted his opponents before the fight began.

Ultimately, Richthofen became an excellent flyer and a fine shot, having always done a lot of big game hunting. But whereas many pilots flew with a kind of innocent courage which had its special kind of magnificence, Richthofen flew with his brains, and made his ability serve him. Analyzing every problem of aerial combat, he reduced chance to the minimum. In the beginning his victories were easy. Picking out an observation plane, he dived on it from the unprotected rear, opened up with a burst and completed the job almost before the enemy pilots were aware of trouble. It was something of this machine-like perfection which accounts for his near death in 1917 after his fifty-seventh victory. Richthofen himself has described the experience:

'On a very fine day, July 6, 1917, I was scouting with my

gentlemen. We had flown for quite a while between Ypres and Armentières without getting into contact with the enemy.

'Then I saw a formation on the other side and thought immediately, these fellows want to fly over . . . We had an unfavorable wind – that is, it came from the east. I watched them fly some distance behind our lines. Then I cut off their retreat. They were again my dear friends, the Big Vickers . . . The observer sits in front. . . .

'My opponent turned and accepted the fight, but at such a distance that one could hardly call it a real air fight. I had not even prepared my gun for firing, for there was lots of time before I could begin to fight. Then I saw the enemy's observer, probably from sheer excitement, open fire. I let him shoot, for, at a distance of 300 yards or more, the best marksmanship is helpless. One does not hit one's target at such a distance.

'Now he flies toward me, and I hope that I will succeed in getting behind him and opening fire.

'Suddenly, something strikes me in the head. For a moment, my whole body is paralyzed. My arms hang down limply beside me; my legs flop loosely beyond my control. The worst was that a nerve leading to my eyes had been paralyzed and I was completely blind.

'I feel my machine tumbling down – falling. At the moment, the idea struck me. "This is how it feels when one is shot down to his death." Any moment I wait for my wings to break off. I am alone in my bus. I don't lose my senses for a moment.

'Soon I regain power over my arms and legs, so that I grip the wheel. Mechanically, I cut off the motor, but what good does that do? One can't fly without sight. I forced my eyes open – tore off my goggles – but even then I could not see the sun. I was completely blind. The seconds seemed like eternities. I noticed I was still falling.

'From time to time, my machine had caught itself, but only to slip off again. At the beginning, I had been at a height of 4,000 yards, and now I must have fallen at least 2,000 or 3,000 yards. I concentrated all my energy and said to myself, "I must see – I must – I must see."

'Whether my energy helped me in this case, I do not know. At any rate, suddenly I could discern black-and-white spots, and more and more I regained my eyesight. I looked into the sun – could stare straight into it without having the least pains. It seemed as though I was looking through thick black goggles.

'Again I caught the machine and brought it into a normal position and continued gliding down. Nothing but shell holes below me. A big block of forest came before my vision and I recognized that I was within our lines.

'If the Englishman had followed me, he could have brought me down without difficulty but, thanks to God, my comrades protected me. At the beginning, they couldn't understand my fall.

'I wanted to land immediately, for I didn't know how long I could keep up consciousness. . . .

'I noticed that my strength was leaving me and that everything was turning black before my eyes. Now it was high time.

'I landed my machine without any particular difficulties, tore down a few telephone wires, which I didn't mind at the moment. . . . I tumbled out of the machine and could not rise again. . . .

'I had quite a good-sized hole – a wound of about ten centimeters in length. At one spot, as big as a dollar, the bare white skull bone lay exposed. My thick Richthofen skull had proved itself bullet proof.'

The bad news of his fall was kept from the German public which superstitiously regarded him as a superman, beyond death. It was less than a month before he was back in the air again, but never his old self. Something had gone out of him. 'Manfred was changed after he received his wounds,' his mother is reported to have said. Now he knew that death could reach him as well as the others, and that is no knowledge for an airman to live with, day and night.

The Richthofen 'circus,' as the Allies called it, was known in Germany as the *Jagdgeschwader*, composed of four *staffels* of five planes each. Toward the end of the War, there were three of these, and their size increased to forty-eight planes. They moved back and forth along the lines from July, 1917, on, wherever the fighting was thickest. It was with *Jagdstaffel* II, Boelcke's old group to whose command Richthofen succeeded, that the greatest German ace gained his long list of victories before the formation of the 'circus'. The Allied planes were camouflaged in colours, but as if in direct challenge, Richthofen's circus was brighter than the sun in colour. His own plane was red from propeller to tail, and the planes of his particular *staffel* were red in kind, with little distinguishing marks, such as a blue tail, white rudder, black aileron, to set them apart from the Red Knight.

For three weeks I lived with the Richthofen *Jagdstaffel*, located at the time on the Ypres front. Ten or twelve officers were living together in a pretty little Belgian country place. This was only a short time before Richthofen was killed, when he commanded the circus and had a great deal of executive work to attend to as well as his daily fighting. Secretaries raced about, and orderlies came and went all day.

Artillery sites were only about fifteen kilometers behind the Front lines, and so, when the circus was scheduled to go aloft, I would start an hour or so ahead of time for the artillery camp, and follow the air fights through their powerful range finders. As a rule the fights would not be more than nine or ten miles off, and two or three miles in the sky.

Spending hours at the artillery range, I saw battle after battle in the air. *Staffel* after *staffel* would leave its airport, circle for height, proceed to the appointed rendezvous in the sky, and form the 'circus' before cruising along the Front in search of Allied squadrons. Richthofen would be flying out in front, the lowest plane in an echelon of Vs, like a flock of immobile geese, fantastically coloured and flashing like mirrors in the sun.

Out of the western skies would come a tinier V of Allied planes, then another and another, until the whole line of them closed with the 'circus' and the blue sky was etched with streaking flight. Round and round, diving, zooming, looping, with motors roaring full out, these lethal wasps spat flaming death through the glittering propeller's disk. Cometlike projectiles missed each other by inches in the whirlpool of sound and fury. Suddenly, out of nowhere, two planes in 125-mile-an-hour flight rushed at each other too late to loop, dive, swerve. Crash! They merged, tangling wings, clasping each other like friends long separated, before gravity pulled them reluctantly apart and they began a crazy descent to bury themselves eight feet in earth miles below. Perhaps I alone noticed them. The taut pilots in the dogfights were taking in sensations with express train speed – flying –

fighting – automatons at the highest pitch of skill and nerve in a frenzy of killing.

Richthofen gained the tail of an enemy. The tracer bullets were spelling out death, when the enemy's engine stopped, the plane went into a quick spin, and only levelled out for a landing quite close to where we were watching the whole battle. We quickly motored over. Richthofen had already gone back to the Front, after landing first, and shaking hands with the officer he had brought down. A bullet had pierced the officer's pocket, ruined a package of cigarettes, travelled on down through his sleeve, punctured his Sam Browne belt and gone on without injury. We looked over his coat, that might so easily have been his shroud.

Asking him to ride with us, we took him back to the flying field, where we picked up Richthofen and together went to the Casino for a good breakfast and friendly chat. I took moving pictures of the officer and Richthofen. Later I acquired a patch of the fabric from Richthofen's sixtieth victory. After a pleasant breakfast, we turned the prisoner over to headquarters, since it was against regulations to keep him for any length of time.

For several days we followed Richthofen's fights. Many of his victories were easy, especially when he attacked the clumsy two-seaters. His usual technique was to dive in their rear, zoom under the tail, and shoot them from very close range. By this time he had become a first class pilot and handled his plane with utmost skill. Seldom did he use more than a quarter of his ammunition on an enemy. Four hundred rounds were carried for each of the two guns. When pilots went from one combat to another, they usually fought until their ammunition was exhausted before returning home.

I think one of the reasons Richthofen survived so long was his ability to keep guarding himself while he attacked. Many other aces were shot down during a fight unexpectedly, as they were training their guns on an enemy pilot. Richthofen would fight very close to his wing men, and not until it was a real dogfight, with the whole air in confusion, would he release his formation to permit every pilot to shift for himself. He was an excellent teacher, and young pilots who showed exceptional skill and courage were sent to his *staffel* to get experience. At first they were taken along to observe the fighting from a distance, and forbidden to engage in combat at all during the first three flights. For it was found that many of the new pilots were killed in their first fight, before they had learned to be all eyes in every direction.

Immediately after each battle, Richthofen would gather his officers for conference and a discussion of the tactics. Occasionally he would censure pilots too aggressive, or too willing to pull away before the battle was over. He was perhaps not so much liked as admired, but the respect other pilots had for him was unbounded.

Proud though he was, the réclame of his feats gave him no particular pleasure. He was not interested in publicity, and though he received letters by the ton from all sorts of people, he cared little for fan mail. When he was around, parties were never wild, for the other pilots felt constrained in the presence of their chief.

Richthofen knew little or nothing about the technical details of airplanes. Unlike Boelcke and Immelmann, he was not even interested, except as it was necessary for him to know for his own safety and development.

While they were alive, we did our best to show the flyers a gay time. It was an open secret that all airplane manufacturers entertained lavishly while the pilots were on leave, and when the aces came to Berlin for the periodical competitions. Because of the popularity of the Fokker plane at the Front, many of the pilots on furlough preferred to make their headquarters with us at the Hotel Bristol. I had a deep admiration for them, and counted many as close friends. Some were so young, I felt almost paternal towards them, although I was only twenty-eight when the War ended.

It was a pleasure to keep open house for the pilots. Naturally it served our interests to hear them talk, discuss one plane and another, the latest tactics of the Allied airmen, sketch their ideal of a combat ship. But what they wanted most, and what we tried to give them was gaiety, charm, diversion, the society of pretty girls, the kind of a good time they had been dreaming about during their nightmare stay at the Front. Berlin was full of girls eager to provide this companionship, for aviators in Germany as in every other country were the heroes of the hour, and the spirit was in the air to make these men happy before they returned to face death alone.

Eddie Rickenbacker was born in a poorish family in Ohio, but thereafter pursued a bruising and exciting life to become one of the classic American heroes. He was one of the earliest racing drivers, and he actually owned and ran the Indianapolis Speedway for twenty years. He became the ace of aces of the US Air Service in France in 1918, with twenty-six kills, and he commanded the famous 'Hat-in-the-Ring' squadron, largely against von Richthofen's 'Flying Circus'. After the war, he concentrated on his own automobile business before taking over and building-up the fledgling Eastern Airlines. In 1941 he made a miraculous recovery from a dreadful crash of a DC-3 in which he was a passenger; a year later he spent twenty-four days adrift in the Pacific on a raft after another crash.

In this extract he describes his memories of the last days of the First World War.

11

FROM: *Rickenbacker*
an autobiography
Prentice Hall, New Jersey, 1967

Those were hectic days. I put in six or seven hours of flying time each day. I would come down, gulp a couple of cups of coffee while the mechanics refueled the plane and patched the bullet holes and take off again. I caught an unguarded balloon while returning from a night mission, and Reed Chambers and I together brought down a Hanover. With the dead pilot at the controls, it glided to a perfect landing two miles within our own lines. We hurried to claim it and had it hauled back to our own field. Then Reed and I each dropped a Fokker in the same dogfight. I shot down a German plane so far behind the lines that the victory was never confirmed. Our 94th squadron pulled out well ahead of the 27th, and after that our lead was never threatened.

In my 134 air battles, my narrowest escape came at a time when I was fretting over the lack of action. I was out alone one afternoon, looking for anything to shoot at. There was a thick haze over the valley of the Meuse, however, and the Germans had pulled down their balloons. To the south the weather seemed a little better; the American balloons were still up. German planes rarely came over late in the afternoon, and everyone had relaxed his vigilance. As I was flying toward the nearest Allied balloon, I saw it burst into flames. A German plane had obviously made a successful attack. Because of the bend in the lines of the front at that point, I saw that I could cut off the Boche on his return to his own territory. I had the altitude on him and, consequently, a superior position. I headed confidently to our rendezvous.

Guns began barking behind me, and sizzling tracers zipped by my head. I was taken completely by surprise. At least two planes were on my tail. They had me cold. They had probably been watching me for several minutes and planning this whole thing.

They would expect me to dive. Instead I twisted upward in a corkscrew path called a 'chandelle'. I guessed right. As I went up, my two attackers came down, near enough for me to see their faces. I also saw the red noses on those Fokkers. I was up against the Flying Circus again.

I had outwitted them. Two more red noses were sitting above me on the chance that I might just do the unexpected.

Any time one plane is up against four and the four are flown by pilots of such caliber, the smart thing to do is to get away from there. There is an old saying that it's no disgrace to run if you are scared.

I zigzagged and sideslipped, but the two planes on top of me hung on, and the two underneath remained there. They were daring me to attack, in which case the two above would be on my tail in seconds. They were blocking me from making a dash for home. I was easy meat sandwiched between two pairs of experts. Sooner or later one would spot an opening and close in.

For a split second one of the Fokkers beneath me became vulnerable. I instantly tipped over, pulled back the throttle and dived on him. As my nose came down I fired a burst ahead of him. Perhaps he did not see the string of bullets. At any rate, he flew right into them. Several must have passed through his body. An incendiary hit his gas tank, and in seconds a flaming Fokker was earthbound.

If I had been either of the two Fokkers above me in such a situation, I would have been on my tail at that very moment. I pulled the stick back in a loop and came over in a renversement, and there they were. Before I could come close enough to shoot, they turned and fled. I suppose that the sight of that blazing plane took some of the fight out of them.

It did not take any fight out of me. I started chasing all three of them back into Germany. We were already three miles behind the lines, but I was annoyed – with them and with myself.

My Spad was faster. One Fokker began to fall behind. He tried a shallow dive to gain speed, but I continued to close in. We were only about a thousand feet up. He began stunting, but I stuck with him and fired a burst of about two hundred shots. He nosed over and crashed. I watched him hit.

All around the crashed plane, I saw flashes of fire and smoke. I was only about five hundred feet above the deck, and the Germans on the ground were shooting at me with all the weapons they had. I could see their white faces above the flashes. The air around me must have been full of flying objects. I got out of there fast and went home to report that I had blundered into a trap and had come out of it with two victories. I now had nineteen.

During the month of October the fortunes of war shifted both on the ground and in the air. From the air we could see the German ground forces retreating, sometimes in complete disorganization. Our bombers were carrying the fight into Germany, and large numbers of German fighters were pulled back from the front in an effort to protect the civilian population.

All along the lines the feeling was growing that the war was coming to an end. I took a three-day leave in Paris and, for the first time, found the streets illuminated at night and unrestrained gaiety.

During the month of October, I shot down fourteen enemy aircraft. On the 30th I got my twenty-fifth and twenty-sixth victories, my last of the war. My title 'American Ace of Aces' was undisputed. The last victory for the 94th Squadron came on November 10. The Hat-in-the-Ring Squadron downed sixty-nine Boche planes, more than any other American unit.

On the night of the 10th a group of us was discussing the next day's mission when the phone rang. An almost hysterical voice shouted the news in my ear: at 11:00 the following morning, the war would end. Our mission was called off. For us the war ended at that moment.

I dropped the phone and turned to face my pilots. Everyone sensed the importance of that phone call. There was total silence in the room.

'The war is over!' I shouted. At that moment the anti-aircraft battalion that ringed our field fired off a salvo that rocked the building. We all went a little mad. Shouting and screaming like crazy men, we ran to get whatever firearms we had, including flare pistols, and began blasting up into the sky. It was already bright up there. As far as we could see the sky was filled with exploding shells and rockets, star shells, parachute flares, streams of Very lights and searchlights tracing crazy patterns. Machine guns hammered; big guns boomed. What a night!

A group of men came out of the hangar, rolling barrels of gasoline in front of them. Perhaps I should have made an effort to stop them, but instead I ran over and helped. We dumped them in an open place, and I struck the match myself. Up roared a bonfire that could be seen for miles. We danced around that blazing pyre screaming, shouting and beating one another on the back. One pilot kept shouting over and over and over, 'I've lived through the war, I've lived through the war!'

Somebody emptied every bottle of liquor he could find into a huge kettle, and the orderlies served it in coffee cups, including themselves in. For months these twenty combat pilots had been living at the peak of nervous energy, the total meaning of their lives to kill or be killed. Now this tension exploded like the guns blasting around us.

We all ran over to the 95th Squadron. They had a piano, and somebody sat down and began banging the keys. We began dancing or simply jumping up and down. Somebody slipped and fell, and everyone else fell on him, piling up in a pyramid. A volunteer band started playing in the area outside. We ran outside again to continue our dancing and jumping and shrieking under the canopy of bursting rockets. Again somebody went down, and again we all piled on and made a human pyramid, this time bigger and better and muddier, a monument to the incredible fact that we had lived until now and were going to live again tomorrow.

In the morning orders came down that all pilots should stay on the ground. It was a muggy, foggy day. About 10:00 I sauntered out to the hangar and casually told my mechanics to take the plane out on the line and warm it up to test the engines. Without announcing my plans to anyone, I climbed into the plane and took off. Under the low ceiling I hedgehopped towards the front. I arrived over Verdun at 10:45 and proceeded on toward Conflans, flying over no-man's-land. I was at less than five hundred feet. I could see both Germans and Americans crouching in their trenches, peering over with every intention of killing any man who revealed himself on the other side. From time to time ahead of me on the German side I saw a burst of flame, and I knew that they were firing at me. Back at the field later I found bullet holes in my ship.

I glanced at my watch. One minute to 11:00, thirty seconds, fifteen. And then it was 11:00 a.m., the eleventh hour of the eleventh day of the eleventh month. I was the only audience for the greatest show ever presented. On both sides of no-man's-land, the trenches erupted. Brown-uniformed men poured out of the American trenches, gray-green uniforms out of the German. From my observer's seat overhead, I watched them throw their helmets in the air, discard their guns, wave their hands. Then all up and down the front, the two groups of men began edging toward each other across no-man's-land. Seconds before they had been willing to shoot each other; now they came forward. Hesitantly at first, then more quickly, each group approached the other.

Suddenly gray uniforms mixed with brown. I could see them hugging each other, dancing, jumping. Americans were passing out cigarettes and chocolate. I flew up to the French sector. There it was even more incredible. After four years of slaughter and hatred, they were not only hugging each other but kissing each other on both cheeks as well.

Star shells, rockets and flares began to go up, and I turned my ship toward the field. The war was over.

Aviators in the USA in the twenties were mostly wild men: air racers, liquor smugglers, barnstormers, stunt flyers. There were no government regulations nor flying licences, and indeed no airlines of any really respectable kind. One could buy a war-surplus Jenny or Standard army training plane (often still packed in its manufacturer's crate) for as little as $50.

Here is an excerpt from one barnstormer's memoirs from those days when aeroplanes and pilots were cheap, when 'a pilot was as good as he talked', as the saying went, 'and some of those gipsies were first-rate talkers'.

12

FROM: *Old Soggy No. 1*

BY Hart Stilwell and Slats Rodgers
Simon & Schuster, New York, 1954

It was Sunday and a cold day, back in 1924. I was hauling passengers at a field – and when I say field I mean farm, not airfield – close to the little Texas town of Van Alstyne. I was charging the customers five bucks for a five-minute ride. Any time some smart customer started counting time on me I looped him. He stopped counting time.

That was before they got to liking loops.

I was making money faster than the gentlemen in Harding's cabinet. But I was cold and hungry and thirsty, and completely out of the only cure for all that – whisky. Preachers and a lot of other good folks around there paying good money to get hauled – no place for a bootleg-whisky breath.

There was a young fellow there who had gone up three times already and spent all his money. He kept hanging around, though, and late in the day he came up to me and wanted to ride again.

'What you going to use for money?' I asked him.

'I don't have any,' he said, 'but I swiped a pint of whisky from Dad's drugstore. I'll give you that.'

'Boy, get in here quick before somebody else does,' I said. There is a time for preachers and a time for whisky. It was whisky time.

The boy climbed in. 'I want you to dive at the crowd and play rough like you do at Love Field,' he said.

'What do you know about Love Field?' I asked him.

'Oh, I was up there once, Saw you crack up.'

I didn't say any more. He was at Love Field one time and saw me crack up. Fine average.

We took off. I pulled up to about 300 feet and made a dive at the crowd. It was every crowd for itself in those days – no CAA to protect the land bugs. I pulled out, but I didn't pull out high enough. Probably had my mind on that whisky. I hit the top of a telephone pole and off went the landing gear, wheels and all.

The boy saw the wheels flying through the air and he turned white around the gill flaps. Right away he figured the wheels were off my ship – smart boy, that one. He didn't want to dive at the crowd any more.

I pulled up to 500 or 600 feet and motioned for him to give me the whisky. I couldn't get the stopper out with my hands, so I got out my knife and worked the stopper out. It fell in the bottom of the ship and I couldn't find it. So there I was, no landing gear on the ship, no stopper in the whisky bottle. Trouble always came to me in wads, like chewing tobacco.

I took a small drink, about a fourth of a pint, and set the bottle down on the seat beside me and started studying whether to crash land it there or fly it back to Dallas and go in at Love Field where I was used to crash landing. Going in at Love Field without any landing gear would be kind of like Old-Home Week. Word would get around and the whole gang, the Lunatics of Love Field, would be out there to cheer me when I crashed. Then we'd get together and throw one hell of a party to celebrate. Soothing my wounds is what they called it.

Then if I went in at Love Field I would have plenty of time on the way to drink the whisky, and I wouldn't be worrying about losing any of it, the way I was while I circled over that chunk of cow pasture.

But it would be dark when I got to Dallas. Not so good. Anyway, there was an ambulance and a fire truck on the ground below me, and there weren't any photographers from the Dallas *Times-Herald* there, so I figured I might as well go on down. It would be nice crashing once without the *Times-Herald* getting a picture of me crawling out of the remains.

I went up to 2,000 feet and cut the motor so it would cool off some by the time I got down. Not so much chance of burning that way. I lined up with the field, got a good grip on the whisky bottle, and came on in. She hit the ground, skidded a little ways, then over on her back she went.

The ship was ruined, which wasn't so bad as you might figure since it was an old World War I Jenny, and I could make enough money in a day hauling passengers to buy another one. The boy and I and the whisky came through just fine. Didn't spill a drop of whisky.

When you've walked away from as many crack-ups as I have, you learn to save the whisky. Or maybe you save two little girls like I did another time when I had to take the wings off between two houses. Mainly you learn how to save yourself – how to keep the motor out of your eye.

I asked a fellow standing there by the ship how much he would charge me to haul the pieces off the field. He said ten bucks. I threw a match on the ship and it burned to the ground right away. Then I asked the man how much he would charge me, and he said five bucks. Five bucks for a match – nice profit.

I gave him the money and went to Dallas, got another ship, and went back up there the next Sunday and hauled everybody in town, preachers and all. And that boy showed up, too. But when he came up close to the ship, I said, 'Look, boy, you get in here and go for a ride and it's no money and no whisky either.'

We dived at the crowd and scattered hell out of them. That time I pulled up high enough.

You didn't have any trouble getting women in those days if you had a ship. They were out at the field waiting for you, and you could pretty much take your pick. I took my pick. They were part of the game.

I guess women haven't changed much. I wouldn't know exactly. But a man in an airplane now is just a man in an airplane, about like a man in an automobile. He doesn't have a chance against an oilman or some fellow with a herd of white-faced steers.

One thing hasn't changed, though. That's the air. When I get up there with a couple of good cloth wings holding me up, I feel just like I did way back in 1912 when I got shed of the ground for the first time and took to the air in that ship I built, Old Soggy No. 1. It was the first ship built in Texas.

I still feel just dandy up in the air if I'm in a ship that lets me do some feeling – not one of those big metal boxcars where you're all locked in. I feel as free as the cowhand of the Old West felt when he was in the saddle, moving up the trail. About the only difference is he could hear himself sing and I can't. But I never was much on singing anyway, and I couldn't do any good at all while flying because I always had a cud of chewing tobacco in my mouth.

So I keep on flying. But when I fly now, I'm mostly looking backward. If you want to look backward with me, why, come along. You'll see a lot of things that will make you think I was flying at a time when all the people were half-crazy. And you'll see some more things that will make you know for sure I was plumb crazy.

I guess I was. I know I never was exactly the same after I started building that ship back in 1912. I was a Barnstormer of the Skies – a breed that's gone.

Charles Lindbergh came from the world of barnstorming and flying the mail, though he was ever more thoughtful than some of those cowboys, but no whit less brave. His solo flight from New York to Paris in 33½ hours in 1927 set the world afire with admiration, made flying respectable, and convinced the world that air travel might yet become something to be taken seriously. The flight was a heroic adventure that is still fascinating today, for Lindbergh effectively mounted the whole operation single-handed – he organized finance, found a suitable plane, re-designed it to match his needs, planned everything himself.

He waited thirty years to publish his full account of it, in a book that is one of the finest in the whole literature of flying. In it he emerges as a warm-hearted mystic, an imaginative dreamer; and in the second excerpt quoted here, there is a hint of the love of animals and obsession with ecology that was to fill – most fruitfully – the latter part of his life.

13

FROM: *The Spirit of St Louis*
BY Charles Lindbergh
John Murray, London, 1957

While I'm staring at the instruments, during an unearthly age of time, both conscious and asleep, the fuselage behind me becomes filled with ghostly presences – vaguely outlined forms, transparent, moving, riding weightless with me in the plane. I feel no surprise at their coming. There's no suddenness to their appearance. Without turning my head, I see them as clearly as though in my normal field of vision. There's no limit to my sight – my skull is one great eye, seeing everywhere at once.

These phantoms speak with human voices – friendly, vapor-like shapes, without substance, able to vanish or appear at will, to pass in and out through the walls of the fuselage as though no walls were there. Now, many are crowded behind me. Now, only a few remain. First one and then another presses forward to my shoulder to speak above the engine's noise, and then draws back among the group behind. At times, voices come out of the air itself, clear yet far away, traveling through distances that can't be measured by the scale of human miles; familiar voices, conversing and advising on my flight, discussing problems of my navigation, reassuring me, giving me messages of importance unattainable in ordinary life.

Apprehension spreads over time and space until their old meanings disappear. I'm not conscious of time's direction. Figures of miles from New York and miles to Paris lose their interest. All sense of substance leaves. There's no longer weight to my body, no longer hardness to the stick. The feeling of flesh is gone. I become independent of physical laws – of food, of shelter, of life. I'm almost one with these vaporlike forms behind me, less tangible than air, universal as ether. I'm still attached to life; they, not at all; but at any moment some thin band may snap and there'll be no difference between us.

The spirits have no rigid bodies, yet they remain human in outline form – emanations from the experience of ages, inhabitants of a universe closed to mortal men. I'm on the border line of life and a greater realm beyond, as though caught in the field of gravitation between two planets, acted on by forces I can't control, forces too weak to be measured by any means at my command, yet representing powers incomparably stronger than I've ever known.

I realize that values are changing both within and without my mind. For twenty-five years, it's been surrounded by solid walls of bone, not perceiving the limitless expanse, the immortal existence that lies outside. Is this death? Am I crossing the bridge which one sees only in last, departing moments? Am I already beyond the point from which I can bring my vision back to earth and men? Death no longer seems the final end it used to be, but rather the entrance to a new and free existence which includes all space, all time.

Am I now more man or spirit? Will I fly my airplane on to Europe and live in flesh as I have before, feeling hunger, pain, and cold, or am I about to join these ghostly forms, become a consciousness in space, all-seeing, all-knowing, unhampered by materialistic fetters of the world?

At another time I'd be startled by these visions; but on this fantastic flight, I'm so far separated from the earthly life I know that I accept whatever circumstances may come. In fact, these emissaries from a spirit world are quite in keeping with the night and day. They're neither intruders nor strangers. It's more like a gathering of family and friends after years of separation, as though I've known all of them before in some past incarnation. They're as different from men, and yet as similar, as the night's cloud mountains were to the Rockies of the West. They belong with the towering thunderheads and moonlit corridors of sky. Did they board my plane, unseen, as I flew between the temple's pillars? Have they ridden with me through sunrise, into day? What strange connection exists between us? If they're so concerned with my welfare, why didn't they introduce themselves before?

I live in the past, the present, and the future, here and in different places, all at once. Around me are old associations, bygone friendships, voices from ancestrally distant times. Vistas open up before me as changing as those between the clouds I pass. I'm flying a plane over the Atlantic Ocean; but I'm also living in years now far away.

Gray scales appear below, vague and misty. I nose down. But fog closes in before I drop a hundred feet. Were those waves real, or did I see a mirage in the mist? I'm not sure. I decide to fly at a thousand feet instead of fifteen hundred. There, I'll have a better chance of making contact with the water.

But I catch only tantalizing glimpses of the sea. Finally I give up searching for it, and resign myself to the cockpit, the instruments, and the strange passengers I carry. In them are solitude and companionship, proximity and distance, a call to death, a guidance to life. One or two, more prominent than the others, ride just behind my shoulder, close but never touching; communicating sometimes by voice and sometimes without the need of speaking.

Mist lightens – the *Spirit of St. Louis* bursts into brilliant sunlight, dazzling to fog-accustomed eyes – a blue sky – sparkling whitecaps. The ocean is not so wild and spray-lashed. It's less ragged with streaks of foam. The wind's strength has decreased, and it has shifted toward my tail.

The plane's shadow rushes in to meet me as I nose down closer to the waves. I last saw it centered in the rainbow, high up in morning clouds. Such a small shadow, skipping from crest to crest, all but losing itself in the troughs, seemingly fearful it won't catch up before I reach the surface.

Brilliant light, opening sky, and clarity of waves fill me with hope. I've probably passed through the great body of the storm. Clouds still lie ahead and on each side – some as fog on the water – some high above. But there are channels of clear air between. Not that I'll follow those channels, but future periods of blind flying should be shorter, and broken up by similar gemlike vistas of the sea. I'm free of the instruments. I can look around again. The gravitation of life is strong.

Is there something alive down there under my wing? I thought I saw a dark object moving through the water. I search the surface, afraid to hope, lest I lose confidence in vision. Was it a large fish, or were my eyes deceiving me? After the fog islands and the phantoms, I no longer trust my senses. The *Spirit of St. Louis* itself might fade away without causing me great surprise. But . . . yes; there it is again, slightly behind me now, a porpoise – the first living thing I've seen since Newfoundland. Fin and sleek, black body curve gracefully above the surface and slip down out of sight.

The ocean is as desolate as ever. Yet a complete change has taken place. I feel that I've safely recrossed the bridge to life – broken the strands which have been tugging me toward the universe beyond. Why do I find such joy, such encouragement in the sight of a porpoise? What possible bond can I have with a porpoise hundreds of miles at sea, with a strange creature I've never seen before and will never see again? What is there in that flashing glimpse of hide that means so much to me, that even makes it seem a different ocean? Is it simply that I've been looking so long, and seeing nothing? Is it an omen of land ahead? Or is there some common tie between living things that surmounts even the barrier of species?

This ocean, which for me marks the borderland of death, is filled with life; life that's foreign, yet in some strange way akin; life which welcomes me back from the universe of spirits and makes me part of the earth again. What a kingdom lies under that tossing surface! Numberless animals must be there, hidden from my sight. It's a kingdom closed to man, one he can fly above all day and never recognize. How blind our normal senses are. We look at a star, and see a pin point of light; a forest is a green carpet to a flyer's eye; the ocean, a tossing mass of water. Inner vision requires a night alone above the clouds, the sight of deer in a clearing, the leap of a porpoise far from land.

My eyes sweep over the waves again, and I climb to a hundred feet. How far from the coast do porpoises swim, I wonder? Do they travel all the way across the ocean, or do they stay near shore and fishing banks? In laying plans for the flight, I didn't think about studying salt-water life as a part of navigation. If I look carefully, there may be other things to see. But the evenness of the horizon is unbroken by ship or sail or smoke. Scan the surface as I may, I find no second spark of life.

Can it be that the porpoise was imaginary too, a part of this strange, living dream, like the fuselage's phantoms and the islands which faded into mist? Yet I know there's a difference, a dividing line that still exists between reality and apparition. The porpoise *was* real, like the water itself, like the substance of the cockpit around me, like my face which I can feel when I run my hand across it.

Britain's first labour government under Ramsay MacDonald in the 1920s had, like every British socialist Government since, the concept of 'nationalisation' and state ownership as one of the cornerstones of its dogma. And the idea of giant airships also powerfully appealed to them. In one of the oddest episodes in British political history, MacDonald's government commissioned two airships in parallel: one was to be constructed by its own Air Ministry, and the other by the Vickers engineering group under the leadership of that engineering genius Barnes Wallis.

Chief calculator in the Vickers team was Nevil Shute Norway, later to achieve reknown as a novelist writing under his first two names; in his autobiography he recorded the bitter feelings of rivalry between the two teams building the 'capitalist ship' and the 'socialist ship'. 'The Air Ministry staff at Cardington considered that they were engaged upon a great experiment of national importance, too great to be entrusted to commercial interests . . . it was impossible to suppose that any private company could compete with Cardington.' The Vickers team felt that the Cardington people 'ought, by rights, to have been in gaol for manslaughter' after the crash of an earlier defective airship they had built, the R.38. Shute was horrified when he read the official report of the R.38 disaster: 'I sat stunned, unable to believe the words I was reading . . . It was inexpressibly shocking . . . I remember going to one of my chiefs with the report in my hand to ask him if this could possibly be true. Not only did he confirm it but he pointed out that no one had been sacked over it.' It is evidence of the ill-will between the two project teams that 'in the five years that were to elapse before either airship flew, neither designer visited the other's works, nor did they ever meet or correspond upon the common problems they had to solve.'

The 'socialist ship' was of course the ill-fated and hopelessly unairworthy R.101, which crashed with the loss of almost all on board on its maiden voyage. The R.101's failure also pulled down its successful rival, for the R.100 was ordered to be dismantled and scrapped before it could enter commercial service.

Perhaps today it is hard to appreciate the spell which gas airships once held over men's minds, and the hopes which they seemed then to offer of becoming the great long-range ships of the sky.

Something of their charm and of their Jules-Verne-ship-of-the-future quality comes through in the two excerpts I have chosen from

*Nevil Shute's book: of the R.100's first flight; and of her first long
test voyage. One should also remember how primitive and limited
were the aeroplanes of those times to appreciate what then seemed to
be the airship's golden possibilities.*

14

FROM: *Slide Rule*

BY   Nevil Shute

Heinemann, London, 1954

A delay of nearly a month elapsed between the
completion of the R.100 and her first flight. A flat calm, such as
occurs in anticyclones, is required for handling an airship safely
in or out of her shed. There was only one mooring mast suitable
for these ships in the British Isles, at Cardington, and R.101 was
hanging on it waiting till weather conditions were suitable for her
to be taken into her shed. For three weeks R.101 occupied the
mast before she could be put into her shed; in the December
weather a further week elapsed before conditions became suitable
to take R.100 out of the very narrow shed at Howden.

R.100 was as big as an Atlantic liner; she was 709 feet long and
130 feet in diameter, which is to say that she was fifty feet shorter
than the old *Mauretania* but half again as wide in the beam. When
loaded for flight she was, of course, as light as a feather and as
capable of being swayed by the least puff of wind. So far as I can
remember the Howden shed provided no more than two feet
clearance each side of the ship as she was manhandled out of the
door; she fitted the shed like a cork in a bottle so that we needed a
dead calm to get her out. If the airship programme had continued
it would have been necessary to devise a mechanical means of
controlling the ship to take her in and out of the shed, and we
spent many months in 1930 working on such matters, but at that
time we were dependent upon manpower and calm weather.

The forecast for December 16th was satisfactory, a dead calm
being predicted for dawn. A party of officials came from
Cardington on the previous evening to take part in the first flight,
including Major Scott, who was in charge of all flying operations
of the two ships. I got to the shed at about 3 a.m.; the country

80

roads were choked with motor coaches bringing the handling party of 500 soldiers to the shed. It was a wonderful, moonlight night, clear and frosty, without a cloud or a breath of wind. We opened the doors of the great shed for the last time, slunk into dark corners to keep clear of the reporters, and stayed waiting for the dawn. In the shed the crew were running their engines slowly to warm up.

The first light of dawn came at last, and at 7.15 we got on board through the control car in the growing light, and the ship was finally ballasted up. Then the order was given to walk the ship aft. A centre line had been painted on the floor and extended out on to the aerodrome and plumb bobs were suspended from the bow and stern; keeping her straight in this way the handling party walked her out. It was all over very quickly. Inside the ship we could not see when she was clear of the shed, but a great cheer from the crowd told us when the bow had passed out on to the aerodrome.

There was very little to be done. Major Scott had her walked out to a safe distance from the shed, swung her round to point her away from it, and checked the ballasting again. I am told that her enormous silvery bulk was very beautiful in that misty December dawn. Scott completed his ballasting arrangements and climbed on board.

The take-off was simple. From the control car Booth emptied a half-ton bag of water ballast from the bow and another one from the stern and leaning from one of the windows of the car, he shouted, 'Let her go.' Inside the ship we heard the cheers and saw the ground receding, and set about our job of finding out our mistakes.

She floated up slowly; at 500 feet Booth rang the power car telegraphs for two of our six engines to go slow ahead. As the ship gathered way the elevator coxswain nosed her upwards to about a 1,000 feet and we made a few slow circles over Howden to try out the controls. This was my prime anxiety, of course; the figures had shown that she could easily be steered by hand with these enormous unbalanced rudders, and now was the moment of proof. But I had nothing to worry about; the first turn showed that she was behaving exactly in accordance with the calculations. She was steered by a wheel like a ship's wheel four feet in diameter. When on a straight course at cruising speed it was

81

impossible to put on more than about three degrees of helm, but that reduction in the effective fin area made her stable on a circle of wide radius and she commenced to turn; as she took up the turn the air flow at the tail was altered and it became possible to put on more helm, so that it took half a minute or more to get her into the condition of turning with full helm upon her minimum turning circle. With this method of control it was impossible to strain the ship by putting on full helm while she was in straight flight because the coxswain was not strong enough, and it proved to be simple and satisfactory.

After a short time we left the vicinity of Howden and flew slowly to York; by the time we got there it was apparent that there was nothing wrong with the ship and we could confidently fly her down to Cardington near Bedford. We circled York Minster and the city, and then, with minds comparatively at ease, we set course for Cardington and went to breakfast.

Breakfast that morning was bacon and eggs cooked on board, the first of many pleasant meals upon that ship. We were all a little elated, and the matter of the parachutes did not appreciably damp our spirits. We had fifty parachutes slung up in various parts of the ship ready for instant use, but there were fifty-four of us on board. We made grim little jokes about the game of musical chairs; after the first few flights the parachutes were removed.

At the same time, throughout the flights of R.100 the action to be taken in the event of disaster was always in the background of my mind, and in view of the eventual disaster to R.101 it is interesting to recall a precaution that I took for my personal safety. In an airship, unlike an aeroplane, there is no great danger of death from violent impact with the ground. Although both ships were filled with hydrogen there was no great danger of fire in the air, for any gas escaping would go upwards and out of the top of the ship, remote from any likely source of ignition. The chief danger, I thought, was of fire after hitting the ground, when broken electrical cables could make sparks in the presence of large masses of escaping hydrogen or petrol; in that case the fire would spread instantly. This in fact, is what happened to R.101. It seemed to me that in such a case the only chance for survival would be to jump on to the inside of the outer cover and cut one's way out, and drop down to the ground, and one would have not more than five seconds to do it in. For this reason before the first flight of R.100 I bought a very large clasp knife and sharpened it to a fine point and a razor edge, and I carried this knife unostentatiously in my pocket throughout the flights that the ship made.

We made a quick trip to Cardington by the standards of those days, cruising at about fifty-five miles an hour on four engines only. We made good about seventy miles an hour over the ground with the assistance of a following wind, and reached Cardington in two hours' flight from York. Here a surprise awaited us; we had assumed that there would be little difficulty in landing the ship on to the Air Ministry mooring mast. So many articles had appeared in the Press about the wonders of this new method of handling airships that it came as a surprise to us to find that the experts on this matter were inexpert in the use of their rather complicated apparatus. On this first flight it took three hours to land R.100 to the mast; three times we had to leave the aerodrome and fly a circuit and come in again, and make a fresh attempt to establish the connection between the steel cable dropped from the nose of the ship and the cable from the masthead laid out upon the aerodrome. The mooring system was essentially sound and at the conclusion of the R.100 flights sufficient experience had been gained in the handling of the ship and in the use of the mast equipment to enable a landing to be made to the mast in about forty minutes, but this result was not achieved without the experience of numerous mistakes. And here we touched the fringe of one of the chief dangers of the airship programme; too many experiments were being made at the same time.

We landed at about three in the afternoon. That evening we held a conference and decided to fly again next day, taking advantage of the calm, frosty weather. There were several minor defects to be made good upon the ship during the night. One engine had a leaky cylinder and another was suspected of having run a big end; air was eddying in violently at cruising speed through one of the big outer cover ventilators and blowing the gasbags about; this had to be sealed up. A dynamo engine could not be turned over with the starting handle. We held our gloomy inquest on it in the middle of the night, only to discover after two hours' work that the starting handle itself was seized solid in its bearings and the engine was in perfect order. These were the inevitable teething troubles of any very large aircraft, but they meant much work.

I got back to the ship that night at about nine o'clock. It was a cloudless, moonlit night, and freezing hard. The ship lay at the mast 100 feet above the ground, brilliant silver in the beams of the floodlights. There was a crew on board her in accordance with the routine procedure at the mast; I found them at supper in their mess, content and settling in to their new quarters. There was no heating in the ship when the dynamo engines were stopped, and it was very cold. In the control car I found Booth, dead tired, wearing a Sidcot suit and huddled in a small arm-chair beside the shore telephone, nursing the ship's black kitten as he kept his watch in the brilliance of the floodlights. I went on aft to the power cars, and worked till midnight with the engineers.

We flew again the next day with the intention of doing speed trials. The speed of R.100 at that time was a sore point at Cardington; with all politeness the officials there professed themselves unable to believe our ship to be at least ten miles an hour faster than their own. R.100 was, in fact, the fastest airship that had ever flown at that time, or to this day, for all I know; her full speed was eighty-one miles an hour. We did not reach full speed upon that flight, however. Prowling through the ship in search of trouble somewhere over Kettering I heard a little flapping noise of fabric in the region of the lower rudder, and discovered that a sealing strip across the rudder hinge was coming unstuck. This was not serious, but we did not care to take the ship up to full speed till it had received attention in the shed; we cruised around for a few hours and landed to the mast again in the middle of the afternoon.

The landing on this occasion was a demonstration of the special qualities of the airship. A thin frosty mist hung over everything; from seven or eight hundred feet it was just possible to distinguish the ground immediately below. In these days every modern aid would be required on such an afternoon to guide an aeroplane on to the runway; in those days flying would have been most hazardous. In the airship everything was peaceful and secure. We had radio telephone communication with the mast, and coming up to the aerodrome we slowed to a mere crawl, running on one engine at about ten miles an hour with two other engines ticking over slow astern ready to check her way if anything loomed up ahead of us. In the control car there was time for a little conference between the officers over each movement of the controls, as 'Putting her nose down a bit, isn't she?' 'Think so? We could afford to be fifty feet lower.' 'Starting to show on the fore and aft level, sir – about two degrees nose down.' 'All right. Coxswain, five degrees elevator up.' And so on. During all the flights that R.100 made I do not think that it was ever necessary to make a quick decision in the way that the pilot of an aeroplane has to; there was always time to talk the matter over if it seemed desirable and decide what action should be taken next. It is impossible to describe what a sense of security this freedom from quick decisions gave to one who was accustomed to fly aeroplanes; rightly or wrongly I felt as safe through all the flights that R.100 made as on a large ship.

We put the ship into the shed that evening, 'put her back in the box' as somebody irreverently described the operation. Here she stayed over Christmas while we sorted out her teething troubles, berthed in her shed beside R.101 before she flew; we now had a good opportunity to examine her in her completed state. We found her an amazing piece of work. The finish and workmanship struck us as extraordinarily good, far better than that of our own ship. The design seemed to us almost unbelievably complicated; she seemed to be a ship in which imagination had run riot regardless of the virtue of simplicity and utterly regardless of expense.

All that day we flew above the fog and clouds, in bright sunshine. We suspended a variety of air speed indicators on cables below the ship and worked her up to her full speed, which as I have said was eighty-one miles an hour. We had several of the Cardington officials on board during this flight, who took this result with long faces. We all tried to get them to disclose what the full speed of R.101 was but they turned stuffy and wouldn't tell us; the subject was evidently an awkward one and we had to abandon it.

The value of a full power trial lies in the effect it has in disclosing weaknesses in the outer cover of the ship. In the case of R.100 the outer cover took up a curious wave-like formation when the ship was at full speed, which was not evident at cruising speed. These harmonic deflections ran longitudinally down the ship from bow to stern; they did not move at all. I climbed all over the ship with the riggers between the gasbags and the outer cover while the ship was at full speed to examine the attachments of the cover to its wiring system but I could find no place where the attachments were unduly stressed. Finally we came to the conclusion that the matter was of no importance, and was probably due to large-scale eddies of the air that passed over the hull.

The outer cover was a weakness in both R.100 and R.101. In each ship the policy had been adopted of building with relatively few longitudinal girders compared with previous ships in order that the forces in the girders might be calculated more accurately, and this decision was undoubtedly influenced by the disaster to R.38. A ship with few longitudinals, however, must of necessity have larger unsupported panels of outer cover fabric than a ship with many. In both ships the outer cover was the main weakness. Extended flight trials were to prove that our outer cover on R.100 was just good enough for the service demanded of it, but only just.

R.100 made another flight about January 22nd to investigate the curious deformation of the outer cover at full speed; we flew to Farnborough, where an aircraft from the experimental station came up and flew alongside us to photograph the cover while I was crawling about inside the hull with the riggers. At cruising speed, about seventy miles an hour, the cover was quite normal. The officials at Cardington grumbled a bit about it, inferring that we were selling them a bad airship, but they had no leg to stand on technically and it was decided to let the matter go.

So far as I remember, Burney clinched the matter by pointing out that our contract only called for a top speed of seventy miles an hour; if they wished we would put a stop on the throttles of the engine to prevent the ship from going faster than that, which would make the cover quite all right.

The contract for R.100 called for a final acceptance trial of forty-eight hours' duration, and a demonstration flight to India. This latter flight was changed to a demonstration flight to Canada when the decision was taken to equip R.100 with petrol engines, because it was thought that a flight to the tropics with petrol on board would be too hazardous. It is curious after over twenty years to recall how afraid everyone was of petrol in those days, because since then aeroplanes with petrol engines have done innumerable hours of flying in the tropics, and they don't burst into flames on every flight. I think the truth is that everyone was diesel-minded in those days; it seemed as if the diesel engine for aeroplanes was only just around the corner, with the promise of great fuel economy. It was decided amicably, therefore, that R.101 would make the flight to India since she had diesel engines, and we would fly to Canada instead.

The final acceptance trial of R.100 lasted for fifty-four hours; it was an easy and effortless performance bearing out the old saying of the R.N.A.S. that the only way to get a rest in the airship business is to take the thing into the air and fly it. We left Cardington on the morning of January 27th with twenty-two tons of fuel on board to burn before coming down, and we landed again on the afternoon of the 29th having lived very comfortably in the meantime. The weather from the start was vile. At Cardington it was misty with a moderate wind and signs of rain; we came down to 700 feet over Oxford but saw nothing of the ground. We never flew lower than about 700 feet in R.100 and seldom so low as that; our normal flying height was between 1,500 and 2,000 feet. We passed on to Bristol in very bad weather, cruising at fifty-five miles an hour on three engines.

By the middle of the afternoon we were on the south coast of Cornwall in a wind of over fifty miles an hour; we put on a fourth engine to bring the speed up to sixty-five, being unwilling to increase speed more until we saw how the ship behaved in the rough weather. She answered her controls well, but in this wind she took a long, slow pitching motion of about five degrees each way. This very slow pitching was the only motion that she ever took in bad weather except once in the vicinity of Quebec when she rolled a little in a patch of violently disturbed air. Nobody could ever have been sick in that ship.

By dusk we were at the Channel Islands and heading up Channel to spend the night out over the North Sea. The ship had quite comfortable little two-berth cabins for fifty passengers, and we who had no routine duties went to bed at a normal time. At two o'clock in the morning I woke up, and becoming aware of my responsibilities I went down to the control car. I found them changing watch. On enquiring where we were, I was told that we were passing over Lowestoft. I asked what we had come there for, and I was told that since the wife of Steff, third officer, came from Lowestoft, we had come as a graceful compliment to empty our sewage tank over the town. Steff was killed in R.101; I do not think he would mind if I recall our ribaldry that night. The ship was then thrown into a turmoil because somebody had drunk the captain's cocoa; I'm not sure it wasn't me. Seeing that she was in good hands and not falling to bits, I went back to bed.

I was up again at dawn, to find us crossing the coast at Cromer on our way to London in increasing fog. As we got near the city the normal white fog changed to thick, black, greasy stuff that left great streaks of oily soot upon the outer cover; it wreathed apart once and disclosed the Tower Bridge immediately below us; then it closed down thick and black. We saw nothing more till that evening after tea, when we picked up the lights of a fishing fleet in Tor Bay.

At about eight o'clock, in the darkness, we went out to the Eddystone and using the lighthouse as a centre we did comprehensive turning trials on port and starboard helm for nearly an hour, calculating the radius of the turns by timing ourselves on a complete circuit and getting the circumference of the circle from the air speed. I have often wondered what the lighthouse keepers thought to see the lights of an apparently demented airship going round and round their lonely rock in the dark night.

All night we cruised the Channel. We went up as far as Portland, and out into the Atlantic to the Scillies. At dawn when I came down to the control car we were cruising over Cornwall; in the rising sun the valleys were all filled with mist as if it had been poured out of a bottle, and this mist was white and gold and pink in the clear light, very beautiful. That morning we went up the Bristol Channel at an easy speed; near Bristol we ran into cloud and rain again. We landed to the mast at Cardington in the middle of the afternoon, with the ship very wet but none the worse for her long flight.

I think it was during this flight that I went outside on top of the ship for the first time. R.100 had a little cockpit on top at the extreme bow, forward of the first gasbag and reached by a ladder in the bow compartment; this was for taking navigational sights with a sextant. From this cockpit a walking way ran aft on top of the ship along one of the girders; this was a plywood plank a foot in width with the outer cover stretched over it. To give courage to the inexperienced it had a rope lashed down along it every two feet or so, to serve as a handhold. This walking way ran the whole length of the ship to the rear of the fins, where another hatch was provided behind the last gasbag. The slope of the hull at the bow cockpit was about forty-five degrees, which made the first part a little tricky to climb, and personally I always went from bow to stern because the rush of air pushed you up the first climb and you didn't have to look down. When the ship was cruising at about sixty miles an hour, as soon as you got to the top, or horizontal, part of the hull you were in calm air crawling on your hands and knees; if you knelt up you felt a breeze in your head and shoulders. If you stood up the wind was strong. It was pleasant up there sitting by the fins on a fine sunny day and whenever I went up there I would usually find two or three men sitting by the fins and gossiping. We kept a watch up there in the daylight hours to keep an eye on the outer cover, and the riggers got so used to it that they would walk upright along this little catwalk with their hands in their pockets, leaning against the wind and stepping over my recumbent body as I crawled on hands and knees.

Francis Chichester won the first transatlantic single-handed sailing race in 1960; in 1962, aged sixty, he raced his own time to New York and beat it by seven days. Years before that, he had emigrated to New Zealand at the age of eighteen with only a few pounds in his pocket; and only eight years later, he was making £10,000 a year in property. He returned to England, learned to fly, bought a Moth, made a trip around Europe in it, and then flew solo to Australia – only the second man in history to do this. Then he put his Moth on floats, and made the first solo crossing of the Tasman Sea from New Zealand to Australia, using two tiny islands as staging posts, having evolved a novel form of aerial navigation to find them. At the second of these, Lord Howe Island, his Moth floatplane sank while he rested overnight; almost undaunted, Chichester stayed there rebuilding it, continuing his journey ten weeks later.

His plan had been to fly around the world; eventually, it was abandoned when his aircraft hit a wire strung between two hills in Japan. It was wrecked, and Chichester severely hurt.

The excerpt I have chosen from his autobiography describes his flight from Lord Howe Island to mainland Australia, after he had rebuilt his machine.

15

FROM: *The Lonely Sea and the Sky*

BY   Francis Chichester
Hodder & Stoughton, London, 1964

Fifty-five minutes out, the mountains were still visible 100 miles astern, like two tiny warts on the face of the ocean. It seemed perfect weather, with a tailwind of forty m.p.h. I was nearly a quarter of the way across in one hour. But the wind had backed, and I was ten miles off course to the south as a result. At 160 miles out I had a shock: the engine backfired with a report loud above the roar – a thing it had never done before in full flight. Was there another piece of skin in the carburettor? I sat utterly still, waiting for the final splutter and choke. The engine continued firing. I reached up and tried the starboard magneto; it was all right. I tried the port, and the engine dropped fifty revs, firing roughly and harshly. A defective magneto – the only parts I had not repaired myself! The engine ran harshly for two minutes while I listened intently, then suddenly it broke into an even smooth roar again. Thank God it was not the carburettor! And I did have two magnetos.

By the end of the second hour the wind had backed still more to the north, but still was driving the plane onwards. I had covered 217 miles, nearly half-way, in two hours. Again through the change of wind I had not corrected enough for drift, and was now twenty-five miles off course to the south. Every mile to the south added length to the flight, but it did not seem to make much odds in such perfect conditions. I changed course another ten degrees to northward.

Then came clouds, and I could see that I might be unable to get a sextant shot. I did not worry much at first – there did not seem much to worry about with a target 2,000 miles wide ahead. But the wind was increasing, and backing persistently. At 250 miles out the sky was completely shut off by dull grey, threatening cloud. I spurred myself to make some hurried drift observations; the wind had increased to fifty m.p.h. from the north-east, and the seaplane was forty-three miles off course to the south. With the drift, we were heading obliquely for the receding part of Australia. Forty-three miles off course seemed a lot, and I wished that I had not allowed it to build up so much. I changed course another ten degrees northwards.

At three hours out heavy rain stung my face. The drift to the south was becoming alarming; the wind had backed till it was now right in the north. I changed course another ten degrees to the north, which put the wind dead abeam. I dared not correct the drift more than that, or I should have made a head wind of the gale, which I sensed would destroy my chance of reaching the mainland. The seaplane was drifting forty degrees to the south, and moving half sideways over the water, like a crab. The rain became a downpour – I had forgotten that it could rain so heavily. I kept my head down as much as possible, but the water caught the top of my helmet, streamed down my face and poured down my neck. We seemed to strike a solid wall of water with a crash. I ducked my head, and from the corner of my eye I could see water leaving the trailing edge of the wings in a sheet, to be shattered instantly by the air blast. On either side of my head the water poured into the cockpit in two streams, which were scattered like windblown waterfalls and blew into my face. I was flying blind, as if in a dense cloud of smoke. I throttled back, and began a slanting dive for the water. Panic clutched me: if I got out of control, I would be too low to recover. But panic meant dying like a paralysed rabbit. I remember saying out loud, 'Keep cool! Keep cool! K-e-e-p c-o-o-l!' The seaplane passed through small sudden bumps, which shook it violently. I looked over at the air speed indicator on the strut, but I could not see either the pointer or the figures in the smother of water. I had to make do with the rev indicator; if the revs increased, the dive was steepening. I felt that I had to find the surface of the sea, for I dared not try to climb blind. If I lost control and tried to spin out, the sea would not show up quickly enough to level off. I sat dead still, moving only my eyes from compass to rev indicator to vertically downwards over the cockpit edge. When the engine speed increased, I used the control-stick lightly with finger and thumb to ease up the plane's nose. There was more chance of the seaplane's flying itself level than of my keeping it level, flying blind. Thank God it was rigged true.

At last I saw a dull patch of water below the lower wing rushing up at me. I pushed the throttle, the engine misfired, and failed to pick up. I thrust the lever wide open as I flattened the seaplane out above the water; the engine spluttered, broke into an uneven rattle and backfired intermittently, its roughness shaking the whole plane. But the plane kept up, and lumbered on. I concentrated on flying. The engine continued with an uneven tearing noise. The sea was only visible a plane's length ahead, where it merged in the grey wall of rainwater. I was flying in the centre of a hollow grey globe, with nothing to help me to keep level except the small patch where the globe rested on the sea. I hugged every wave, rising or falling with it, and the seaplane jerked its way along. The water poured over my goggles, distorting my sight; it ran down my face and neck, and streams of it trickled down my chest, stomach and back. I dared not take my eyes off the water to look at the compass or the rev indicator. One thing helped me – the violence of the gale itself. Although the seaplane headed in one direction, it was being blown sideways, so that it crabbed along half left, and I could see the next wave between the wings instead of its being hidden by the fuselage. I steered by the drift, keeping the angle of it constant. Otherwise I should have wandered aimlessly about the sea. There was a furious cross sea. Waves shot upward, to lick at the machine, but were slashed away bodily southwards by the wind. The tails of spume streaking south across the wave troughs enabled me to steer a straight course. I knew that I was flying as I had never flown before, but I also knew that I could not last long at that pace. At any moment I expected a muscle to lag, and the seaplane to strike a wavecrest. Suddenly I found myself flying straight into the water, and snatched back the stick to jump the seaplane's nose up, thinking my eye and hand had at last failed me. Then I realized that the rainfall had eased while I was in the trough between two rollers, and that the crest of the swell ahead had unshrouded above me. Next instant the seaplane shot into the open air. I rose 30 feet, and snatched the goggles up to my forehead. It seemed like 3,000 feet.

The compass showed that I was fifty-five degrees off course, headed to miss even Tasmania. That seemed strange, for I thought that I could fly accurately by the drift. I soon saw what had happened – the wind had backed another forty-five degrees, and was now north-west. I had to think hard. I picked up the chart case on which my chart was rolled, but the soaked chart was useless. Before the storm I had been drifted so far south that I was right on the edge of the chart; during the storm I was blown farther south at the rate of a mile a minute, and was now far off it. I had a small map of Australia torn from a school atlas on the island. There was not enough sea area on it to show Lord Howe Island's position, but it was the only thing to use. Where was I before the storm? The position on the school map came right in the middle of a city plan of Sydney in the corner.

I flew up to another line squall ahead, parallel to the previous storm, and stretching from horizon to horizon, but I could see the water on the other side, as through a gauzy curtain. The rain was heavy, but the engine still carried on. We flew through the curtain of rain into an immense cavern of space between the illimitable vault of dull sky above, and the immeasurable floor of dull water below. It was solitary in that great space. Some slanting pillars of rain leaned against the wind, trailing across the dull floor of water like spirits of the dead drifting from the infernal regions. The vastness lent it all a nightmare air. In one place the vaulted ceiling bulged downwards with two black-cored squall clouds, each linked to the sea by a column of waterspout. Between the two columns another waterspout, a slender grey-white pillar was rising from the sea's surface. At a good height it burst at the top, like smoke expanding after an explosion. I flew straight towards it, fascinated. Suddenly the engine burst into a rough clatter again, and I realized that I must not fly near that thing; the disturbance capable of twisting it from the sea must be terrific.

I thought I saw land away to the north-west, purple foothills with a mountain range behind, but when I looked for it again it had disappeared; land was still 160 miles away. At the foot of a great storm-cloud I saw smoke – a ship. It offered me a new lease of life, and I immediately turned towards it. It lay at the edge of the storm like a duck at the foot of a black cliff. I swooped down and read the name *Kurow* on the stern. It was an awesome sight. The bows slid out of one comber, and crashed into the next, to churn up a huge patch of seething water. When a cross swell struck her, she lurched heavily, slid into a trough and sank, decks awash, as if waterlogged; but wallowed out, rolling first on one beam and then the other, discharging water from her decks as though over a weir. There was no sign of life on board, and I could not imagine anything less capable of helping me. I felt as if a door had been slammed in my face, turned and made off north-west to round the storm. I felt that I would rather go fifty miles out my way than face another storm. I had been only four hours thirty-five minutes in the air: it seemed a lifetime.

Round the storm we flew into calm air under a weak hazy sun. I took out the sextant and got two shots. It took me thirty minutes to work them out, for the engine kept backfiring, and my attention wandered every time it did so. The sight in the end was not much use; the sun was too far west, but I got some self-respect from doing the job.

Suddenly, ahead and thirty degrees to the left, there were bright flashes in several places, like the dazzle of a heliograph. I saw a dull grey-white airship coming towards me. It seemed impossible, but I could have sworn that it *was* an airship, nosing towards me like an oblong pearl. Except for a cloud or two, there was nothing else in the sky. I looked around, sometimes catching a flash or a glint, and turning again to look at the airship I found that it had disappeared. I screwed up my eyes, unable to believe them, and twisted the seaplane this way and that, thinking that the airship must be hidden by a blind spot. Dazzling flashes continued in four or five different places, but I still could not pick out any planes. Then, out of some clouds to my right front, I saw another, or the same, airship advancing. I watched it intently, determined not to look away for a fraction of a second: I'd see what happened to this one, if I had to chase it. It drew steadily closer, until perhaps a mile away, when suddenly it vanished. Then it reappeared, close to where it had vanished: I watched with angry intentness. It drew closer, and I could see the dull gleam of light on its nose and back. It came on, but instead of increasing in size, it diminished as it approached. When quite near, it suddenly became its own ghost – one second I could see through it, and the next it had vanished. I decided that it could only be a diminutive cloud, perfectly shaped like an airship and then dissolving, but it was uncanny that it should exactly resume the same shape after it had once vanished. I turned towards the flashes, but those too, had vanished. All this was many years before anyone spoke of flying saucers. Whatever it was I saw, it seems to have been very much like what people have since claimed to be flying saucers.

I felt intensely lonely, and the feeling of solitude intensified at every fresh sight of 'land', which turned out to be yet one more illusion or delusion by cloud. After six hours and five minutes in the air I saw land again, and it was still there ten minutes later. I still did not quite believe it, but three minutes later I was almost on top of a river winding towards me through dark country. A single hill rose from low land ahead, and a high, black, unfriendly-looking mountain range formed the background. A heavy bank of clouds on top hid the sun, which was about to set.

Well, this was Australia. Away to the south lay a great bay, and at the far side I spotted five ships anchored. They were warships. I flew south, and crossed the bay. Flying low between the two lines of ships I read HMAS *Australia*, HMAS *Canberra*. On the other side there appeared to be an aircraft-carrier. My heart warmed at the thought of getting sanctuary there, but all the ships had a cold, lifeless air about them. I supposed that I must fly on to Sydney. I flew over an artificial breakwater near a suburb of red-bricked, red-tiled, bungalows and houses like a small suburb in a dull-brown desert, with only a few sparse trees of drab green. There was not a sign of life, and not a wisp of smoke from the chimneys. Had the world died in my absence? If there was anyone left alive, there would surely be a watchman on one of the warships. I turned and alighted beside the *Australia*, its huge bulk towering above me. The seaplane drifted past and away from it, bobbing about on the cockling water. There was dead silence except for the soft *chop chop chop* against the float. I felt a fool to drop into this nest of disdainful battleships. I stood on the cockpit edge, and began morsing to the *Canberra* with my handkerchief. An Aldis lamp at once flashed back at me from the interior of the bridge. A motor-launch shot round the bows of the warship. I cancelled my signal, and stood waiting.

'How far is Sydney?'

'Eighty miles.'

I dreaded the thought of Sydney, and its crowds, but my job was to reach it. The launch was crowded with sailors, and at 20 yards their robust personalities gave me a feeling of inferiority. I felt that I had to get away quickly. I asked the launch to tow me to the shelter of the breakwater, and a sailor slipped me a tow rope efficiently. I climbed out on the float to swing the propeller, and as I swung it I noticed mares' tails of sticky black soot on the cowling, due to the backfires. I wondered if the engine still had enough kick to get me away, but as soon as the seaplane started moving forward and pounding the swells, the futility of trying to take off was obvious. That settled it; I had to ask for help. The launch approached again. 'We'll tow you to *Albatross*,' an officer said. I made fast the tow line and I was towed up to the aircraft carrier, where I made fast to a rope dangling from the end of a long boom. I released the pigeons, feeling sorry for them, and they took off flapping and fluttering, presumably for their home loft near Sydney. A sailor let down a rope ladder from the boom, and I grappled clumsily up it, my feet often swinging out higher than my head. I made my way along the boom to the deck where a commanding figure, with much gold braid, was waiting for me. 'Doctor Livingstone, I assume,' he said, looking hard at me. 'At any rate, you have managed to discover the only aircraft carrier in the Southern Hemisphere. Come along to my cabin.'

I felt like a new boy in front of the headmaster. 'Did I say, when you came aboard, "Doctor Livingstone, I *assume*?" Of course, I meant, "Doctor Livingstone, I *presume*?".' But Captain Feakes of the Royal Australian Navy was a great host. He gave me a whisky and soda and made me feel like a long-expected, favourite guest. Yet I felt isolated, and drained of personality, horribly cut off from other people by some queer gulf of loneliness. I had achieved my great ambition, to fly across the Tasman Sea alone, I had found the islands by my own system of navigation which depended on accurate sun-sights worked out while flying alone, something which no one had ever done before and perhaps no one ever would do in similar circumstances. I had not then learned that I would feel an intense depression every time I achieved a great ambition; I had not then discovered that the joy of living comes from action, from making the attempt, from the effort, not from success.

Squadron-Leader Hewitt of the Australian Air Force arrived and offered to lift the seaplane on board *Albatross*. I asked him to let me do the job of hooking on. It was dark when I went on deck. An arc-lamp shed a brilliance high up but only a dim light reached the seaplane as she was towed slowly under the lowered crane-hook. Standing on the top of the engine of the bobbing seaplane. I tried to catch the ponderous hook; it was a giant compared with the one at Norfolk Island, with a great iron hoop round it, probably a help in hooking on big flying-boats, but only adding to my difficulties. I had to duck the hoop to catch the hook with one hand, and reach under it with the other to keep the two sling wires taut with the spreaders in place and the middle points of the wires ready for the hook. The hook itself was so heavy that I could not lift it with my arm outstretched. The seaplane was rolling, and also there was a slight movement of the aircraft carrier, sufficient to tear the hook from my grasp, however tightly I clung with my knees jockeywise to the engine cowling. At last I had the wires taut and the hook in place under them, when either the seaplane dropped or the aircraft carrier rolled unexpectedly. The hook snatched and lifted the seaplane with my fingers between the hook and the wires. The pain was excruciating, as the wires bit through my fingers. I shrieked. I felt ashamed; but I knew that my cry was the quickest signal I could give the winchman. The hook lowered, and I sat on the engine top, knees doubled up, leaning against the petrol tank. I could not bear to look at my hand. The hook swung like a huge pendulum above me. I felt, well I had bragged of my skill at this job; I should just have to get on with it. I cuddled the round of the iron hook in the palm of my right hand, and rested the wires in the crook of my thumb of the other hand. Everything went easily. 'Lift!' I said. The water fell away, and at last the seaplane swung inboard, stopped swinging, and dropped softly on to padded mats. I said to a man standing by, 'Help me down, will you? I am going to faint.'

When I came to I was in the ship's hospital. My right hand was crushed, but I lost only the top of one finger. The surgeon cut off the crushed bone and sewed up the flesh. I then became the guest of the wardroom officers as well as of Captain Feakes, and it is hard to recount such marvellous hospitality. It was like staying in the best club with the mysterious fascination of naval life added.

*Air travel is one of the great business success stories of the century;
the ever-increasing speed and size of transport aircraft have given the
industry enormous increases in productivity, enabling more and more
people to fly at less and less cost – in comparison to their personal
incomes. The speed at which the industry has grown is astonishing: as
late as the 1950s, more people were still crossing the Atlantic by sea
than by air; while now only one liner, the QE2, still sails regularly
between Britain and America. Today, British Airways is a giant
corporation employing 55,000 people including 3,000 pilots; forty-
five years ago its ancestor, Imperial Airways, had only a handful of
aircraft and just thirty-two pilots. Cruise speeds of the rattly old
biplanes of those days were around 100 mph; today, Concorde flies at
1,200 mph.*

*The accomplished American novelist Ernest K. Gann says of air
travel: 'It has lofted the common man towards hitherto undreamed of
destinations, and it has changed forever the very pattern of his daily
existence.' Gann began his long flying career in 1935, barnstorming in
leather jacket and helmet; his point-of-sale display was a canvas sign
that read* AIRPLANE RIDES *$2.*

*The airlines were just then moving into streamlined metal
monoplanes with retractable wheel undercarriages; here, in his most
recent book, Gann looks back to the baroque old biplane flying boats
that preceded them: an account of the Imperial Airways Mediterranean service, then still largely rooted in nautical traditions.*

16

FROM: *Ernest K. Gann's Flying Circus*

Hodder & Stoughton, London, 1976

The late 1800s brought a traumatic time to mariners, for
all those hearties so long accustomed to the glories and miseries of
sail saw their life style change exceedingly with the advent of
steam power. A few stubborn sailors kept their canvas flying
until the very end, but the majority accepted the inevitable doom
of themselves and their vessels and swallowed the anchor with
such grace as they could manage. Those who were realists
adapted their ways to steam and thereby survived.

Perhaps no other aircraft in the flying world offers a better parallel to that era of fading romance than the 'Scipio' class flying boat, which displayed the British flag with considerable hautiness along a sector of Imperial Airways route to the East. In true, empirical style it was a 'pukka sahib' of aircraft and quite as anachronistic. Properly dressed passengers carried their cork helmets handy for the landings in Egypt, and they were *all* 'properly' dressed.

'One must keep one's status clear along the way to Injia. . . .'

Although such quaint national pride may now be long gone with the winds of colonialism and guilt about the white man's burden, it was *de rigueur* for Scipio passengers.

While the British Empire crumbled piece by piece, the drill was to look the other way. It was not happening. Much of this conservative myopia seeped into the operational philosophy of Imperial Airways. Thus, while the Americans were already flying the Boeing 247 and creating the Douglas DC-2, the English were still setting sail in such Victorian geese as the Scipios. And sadly, when one by one they departed the scene, a way of life for aerial sailors was gone forever. So also passed one of England's several chances to dominate the skies and aircraft manufacture as it had once ruled shipbuilding and the seas.

Although commonly known as Scipios, the official designation of these candidates for the world's most rococo aircraft was 'Short S-17, Kent.' The English have always had the decency and taste to name their aircraft types rather than number them in the monotonous American fashion, and so the 'S-17' was ignored along with 'Short,' the manufacturer's name. Somewhere even the 'Kent' became lost, and the type was stuck with the name of the first to fly. Thus Scipio.

The year is 1932. Only three Kents are built, Scipio, Sylvanus, and Satyrus. While their number seems minute in comparison with present quantities it is a direct reflection of the aerial times. Only thirty-two pilots are regularly employed in the whole of British air transport, a number exactly matched by the number of airline transport aircraft. Across the channel, 135 French pilots are flying 269 aircraft, while the Germans somehow manage to fly 177 aircraft with only 160 pilots and still far exceed any rivals in weekly air mileage.

Apparently the production of only three aircraft has some

magic quality about it which applied to flying boats. Built and flying during almost the same period (1931–1934) are Pan American's only three Sikorsky S-40s, a four-engined flying boat hoisting nearly as many aeronautical square sails as the Scipios. Also contemporary are the grandiose DO-Xs which so very clearly demonstrate that even the usually progressive Germans have not been able to put aerodynamics and flying boats together. Only three DO-Xs are ever to be built, which was just as well. Still in the future will be the last of the big four-engined flying boats, the Vought-Sikorsky 44s. Again only three will be built.

All the while Glenn Martin is brooding in Ohio and Maryland. A most peculiar individual, somewhat of a charlatan and somewhat of a genius, he has always been a stout champion of flying boats. With his eventual production of the Martin 'Clippers' plus his very genuine energy and influence he will do much to keep the aeronautical anachronisms flying long past their legitimate time.

The Scipios ply the seas and sky mainly between Brindisi in Italy and Alexandria in Egypt. In so doing they are playing a vital role in aviation history for their very assignment is the direct result of a national philosophy which if carried to equal extreme by additional governments might chain international flights to the ground and cause incredible difficulties for worldwide air transport.

Soaking complacently in a tub of bureaucratic arrogance the French and the Italians have proclaimed sovereignty of the atmosphere above their respective real estate and have forbade any 'over-flight' with passengers. Thus anyone bound for England to the Far East is obliged to go by sea or take a flight from London to Paris, transfer to a train, and proceed overland to the port of Brindisi. There, having paid due tribute to the French and Italian railroads, the passenger may transfer to a Scipio and resume eastward progress. The resulting shambles if every nation in the world should take such bullheaded attitudes could conceivably affect flights into outer space, a fancy which such men as H. G. Wells, Jules Verne, and Robert Goddard have visualized.

Eventually the restrictions are to be abandoned, but not without a gesture from a volatile Italian who will express his continued objection by setting fire to and destroying the flying boat Sylvanus in Brindisi harbour. Rumour will say he is a railway employee.

Like other vehicles of transport all large flying boats were born of a common denominator – an international lack of airports. In the time of Scipios it is ever so much easier to put down a few mooring buoys and hire the local beach boys and their 'bumboats' for connection with the shore than it is to build runways. Men of almost every nationality are familiar with the sea, but very few know anything whatsoever about creating airports. The general conception of a ship that flies, especially if the route is over water, is relatively easy to accept and it will prevail until a still unforeseen world war proves the 'safety' factor of flying boats alighting in anything but protected harbours is hardly much better than a 'ditched' landplane. No one will quite believe that the occasions for unscheduled landings in the briny will be so rare that the chance may be virtually ignored in aircraft design.

While amphibians suffer the worst reputations for expensive maintenance, flying boats have always run a very close second, and both types are cursed with innumerable problems which are not present in less glamorous aircraft. Corrosion is only one persistent battle involved in flying boats, and operation in their liquid element creates an almost unresolvable conflict between aerodynamic efficiency of design and water-born demands. At normal landing speeds water becomes nearly as hard as concrete, and so the hull of a flying boat must be built over-strong, which of course takes a penalty from the useful weight load. Likewise the engines have to be fixed high above any possible spray since mere water droplets can very quickly destroy the efficiency of whirling propeller blades.

While comptrollers are frowning over maintenance costs against revenue earned, the flight operations people of any flying boat transport company are constantly reminded of those hazards peculiar to their crafts, some of which just cannot be anticipated even with the best techniques. 'Glassy water' landings can be damned dangerous if not accomplished just right, and *all* night landings offer more than enough opportunities for accident. Striking a single dead-head log on landing or takeoff can sink a flying boat, and a flat calm day can cause an embarrassing delay in departure while the frustrated pilot chases his tail around trying to create enough wake to break hull suction. Otherwise intelligent surface mariners have always shown a strange indifference to understanding vessels in flight, and crowded harbours can present nightmare effects to a flying boat skipper who is trying to set down between a determined ferry, an odd freighter or so, and various small craft whose helmsmen have apparently been struck blind. And then there is always the over-anxious launchman bringing either officials or supplies alongside who refuses to recognize that the relatively fragile topsides of a flying boat will not repulse his clumsiness with the same ease as a steel ship of thirty thousand tons. As for schedules, realistic flying boat operatives soon acquire a wistful air. Try as they might to eliminate mechanical delays the vagaries of wind and weather have a much greater effect on their aircraft than on landplanes. Even arrivals and departures of the reliable Scipios are occasionally two or three days late.

Yet many amenities soothe those passengers who are in enough haste to travel by flying boat. The Scipios carry only sixteen passengers which allows people a chance to stretch their legs and be treated as honoured guests aboard rather than as so many faceless bodies weighing (with baggage) a uniform 170 pounds. The cuisine is excellent and served on tables covered with lace doilies meticulously squared about a vase of fresh flowers. Spotless napkins are folded in the water glasses to be removed and snapped with gastronomical solemnity by the anticipating diner. And the finest wines are served in cut-crystal goblets – the reds uncorked and allowed to 'breathe' a suitable time before tasting and the whites chilled yet not frozen into a flavourless pseudo-sherbet. Nor is the passengers' peace of mind or conversation constantly shattered by inescapable pronouncements via a public address system. If anyone really cares they simply glance at the airspeed and altitude indicators conveniently placed on a cabin bulkhead. Airspeed 105 miles per hour. Altitude 1,000 feet. The windows are easily available to everyone aboard, and the view from a decent altitude only contributes to an undeniable yet strangely pleasant sense of being somewhat superior to the rest of mankind.

'Ladies and gentlemen, if you will step this way the launch will take you ashore.'

The Scipios provide an elegance, a style, and a drama of self-respect never to be played again. And sometimes if a crocodile chooses to snooze on the mooring buoy urging his departure is also taken for granted. God is in his heaven and Britons, Britons . . . never shall be slaves.

All three Scipios were eventually destroyed. The namesake crashed on landing at Mirabella in Crete, and in 1938, nearly three years after the passionate Italian scuttled Sylvanus, the Satyrus was scrapped.

Two landplane versions of the type were built and dubbed Scylla and Syrinx. They flew the route between London, Paris, and Brussels and were not at all popular with either pilots or maintenance crews. Everything was out of convenient reach for mechanics, and a session at the ponderous controls left the strongest pilots physically weary.

A Frenchman named André Jacques Garnerin made the first parachute descent, from a balloon in 1797. The first descents from aeroplanes were made in America in 1912 and in England in 1914. German military crews were issued parachutes late in the First World War, but Allied aviators were not, lest their will to fight be sapped. Such thinking seems savage and unrealistic today when combat aircraft crews have parachutes as a matter of course. Many thousands of lives have been saved in the last fifty years; there have even been a handful of aviators who have lived after falls without parachutes. (One, a Russian named Lieutenant I. M. Chisov, fell 22,000 feet, $4\frac{1}{2}$ miles, in 1942 into deep snow, escaping with a fractured pelvis and spinal injuries. An RAF Flight Sergeant, Nicholas Alkemade, fell chuteless from a blazing Lancaster over Germany in 1944, landing in a fir tree without a single broken bone.)

Equally strange is the account that follows of a glider pilot testing a new design of sailplane who tried to bail out (when it would not recover from a spin) but failed.

Gliding for sport and soaring in rising air currents had been developed by the Germans in the 1920s, when the terms of the Treaty of Versailles forbade them to build or operate powered aircraft. It was they who discovered that upcurrents near the windward face of hills and in thermal currents on sunny days were easily enough to sustain a glider that was carefully streamlined. This new sport of soaring spread quickly around the world, with contests both national and international being organized so that soaring pilots could pit their skills against each other.

Philip Wills was testing a new British glider for one of these in 1937 when he decided to take to the silk. Though a superbly skilled glider pilot he was no professional test pilot, but rather a businessman for whom gliding was a hobby.

17

FROM: *On Being a Bird*

BY   Philip Wills

Max Parrish, London, 1953

The first machine was finished in April, and I went
north for her test flying, which took place from York aerodrome.
She proved beautiful to look at and handle, and everyone
congratulated everyone else on her success. A month went by,
and in Coronation week I again went north for her final trials, the
last item of which was to be her test for spinning.

Even today, when the causes of spinning and the design
requirements for successful recovery from a spin are fairly well
known, every machine has to be taken to a great height and
stalled and spun deliberately before the designer can be quite
certain that it is satisfactory in this respect. But I had never heard
of a glider spinning viciously: in fact, my own Hjordis refused to
spin at all; and the King Kite had behaved so beautifully on all her
trials so far that any idea of trouble never entered my head.

I was strapped into the cockpit and the celluloid cover was put
on over my head. Three hundred feet of wire cable was strung
from the tail-skid hook of the towing aeroplane to my own nose-
hook, and we took off.

It was a clear blue day with not a ripple in the sky. At 4,500 feet
I pulled the release, and the aeroplane put down its nose and
dived away to the distant earth. I floated along for a minute or so
in the blissful quiet which is one of the abiding joys of the sport.
Then I eased the stick gently back, and she started to climb.
Slowly the speed fell off until, as the needle came back to just under
40, she gave a little shudder, and the stick went dead. I kicked on
full left rudder, the nose rolled over, the earth tilted majestically
up from underneath until it was right ahead of me, then started to
revolve.

After half a turn the speed came up rapidly, I put on opposite
rudder and centralized the stick in the normal way, the earth
slowed down and returned quietly to its normal place beneath
me. Good! I glanced at the altimeter: still 4,000 feet up, plenty of
height for one more spin, the other way.

I eased back the stick once more, she climbed, slowed, faltered. I kicked on full right rudder, she rolled over like a gannet on to the dive. I let her spin a little longer this time, all seemed well, then moved the controls again to bring her out.

Nothing happened: the earth ahead went on revolving like an immense gramophone record, objects on it growing perceptibly larger.

Quickly I put the controls back to spinning position, adjusted the wing flaps, and tried again, firmly. Still the spin went on, my speed increasing unsteadily and the hum of the wind outside growing to a roar. There was nothing more to be done; the time, so often anticipated, had come to abandon ship and take my first taste of the delights or otherwise of parachuting.

There was so much to do that I did not feel in the least worried, only faintly ridiculous that such a thing should befall me, a respectable City business man, husband and father. There was some mistake; these experiences should confine themselves to our professional heroes. But my next movements had been practised so often in imagination that they took place almost without volition.

I reached above my head for the cord releasing the cockpit cover, pulled it, and pushed the cover upwards. The gale outside lanced in underneath as it lifted, caught it from my hand and whirled it away. The wind shrieked and tore at my clothes, caught my glasses and whipped them off. The nearing earth became a green, unfocused, whirling blur. I let go of the controls and my hands went to the catch fastening the safety straps, clicked it open, and gave a reassuring pat to the sheathed handle of the parachute rip-cord.

The machine, freed of all attempt at human control, lurched about as it spun in a curious idiot way, a body without a mind. I drew up my knees, leant over the left-hand side of the cockpit, and dived head foremost over the edge.

And now a dreadful thing happened. As I went over the side, it seemed to swing round and up at me with a vicious jerk, struck me across my chest and flung me back helplessly into my seat. The spin went on.

A second time I gathered myself together and leaped over the side, and a second time I was caught and bounced childishly

107

back: it was as if I was struggling to break through the bars of an invisible but invincible cage. The spin went on. The blurred earth seemed now very near: in fact, it was scarcely two hundred and fifty yards ahead.

It is curious but true that at a last moment such as this one's body, partly exhausted by former struggles, can nevertheless gather sufficient physical energy to make a final effort surpassing previous ones. A third time I flung myself, still more violently, over the side. This time I got well out, head-first and well forward – almost free – the whirling outline of wings and fuselage was all around me, filling the sky. There was the most appalling bang, a violent blow, and I found myself once more back in my seat, hands and feet instinctively on the controls. The world swung violently overhead, slowed, stopped spinning, the machine was on its back, only centrifugal force was holding me in – but the controls were biting the air again, life had come back into them, life in them was life in me. I pulled back on the stick, she came staggering round on the second half of a loop, I steadied her, and looked down.

The aerodrome buildings were a bare three hundred feet beneath. The raging gale in the exposed cockpit sank to a friendly breeze, apologizing for losing its temper. I put down the landing flaps, did a half turn and landed, not a hundred yards from the club-house. I looked out, and saw my wife running towards me over the turf. My recording barograph showed that I had spun down like Satan from Paradise, two-thirds of a mile in one minute.

After a while I levered myself to my feet and stepped gingerly over the edge of the cockpit. As I did so I felt a violent cramping pain in my chest. Clearly I had strained my heart: you can't go doing things like this without paying for it, I thought.

I was helped into the club-house and lay down, waiting for the reaction. Every time I moved, my heart gave a vicious tweak. Ten minutes later I put my hand to my waistcoat pocket for a soothing cigarette and brought out my metal case bent in a line across its centre to the shape of a three-dimensional 'L'.

My last jump had got my weight so far forward, had so materially altered the trim of the machine, that the hidden vice had been overcome. As the wings bit the air at high speed the machine was swung round with such violence that we found the main wing-bolts, great rods of steel holding the wings to the fuselage, all four bent. And the side of the cockpit came round and hit my chest with a blow sufficient to break every rib in my body. But instead, it caught my cigarette case fair in the middle, the force was distributed over an area, and I escaped with a set of internal bruises that kept me awake for the next three weeks, a cigarette case that will never open again, and a parachute that never had been opened, but nevertheless, had given me the incentive to jump.

In the outcome, I was able to describe my spin sufficiently well for the experts to diagnose too small a rudder. A larger one was fitted, the machine was tested again with many safeguards. Although subsequent events showed that this was not the full answer, we took the three King Kites to Germany with two other machines, and in the fortnight the British team did cross-country flights totalling over 1,100 miles.

I was told that I should have known that you must always bale out on the inside of a spin, and that if, in a right-hand spin, I had dived over the right-hand side, I would have fallen cleanly and instantly down the centre of the corkscrew and, once clear of the machine, released my parachute and watched the glider spin down to its wreckage.

As it happened it was fortunate I did not know this, one of the elementary rules of parachuting which no one troubles to tell you; for the machine was saved. Even the cockpit cover was picked up, but slightly damaged, a mile away. The only permanent loss was my spectacles, and perhaps a few days cut off my old age.

There's been an ocean of poetry written about flying – none of it much good – but I've always liked this little (and little-known) gem of W. H. Auden's. No flyer himself, he has somehow got the feel of things here. It is one of the cleverest examples of use of alliteration that you will find.

18

The Airman's Alphabet

FROM: The Orators
BY   W. H. Auden

ACE
Pride of parents
and photographed person
and laughter in leather.

BOMB
Curse from cloud
and coming to crook
and saddest to steeple.

COCKPIT
Soft Seat
and support of soldier
and hold for hero.

DEATH
Award for wildness
and worst in the west
and painful to pilots.

ENGINE
Darling of designers
and dirty dragon
and revolving roarer.

FLYING
Habit of hawks
and unholy hunting
and ghostly journey.

GAUGE
Informer about oil
and important to eye
and graduated glass.

HANGAR
Mansion of machine
and motherly to metal
and house of handshaking.

INSTRUMENT	Dial on dashboard and destroyer of doubt and father of fact.
JOYSTICK	Pivot of power and respondent to pressure and grip for the glove.
KISS	Touch taking off and tenderness in time and firmness on flesh.
LOOPING	Flying folly and feat at fairs and brave to boys.
MECHANIC	Owner of overalls and interested in iron and trusted with tools.
NOSE-DIVE	Nightmare to nerves and needed by no-one and dash toward death.
OBSERVER	Peeper through periscope and peerer at pasture and eye in the air.
PROPELLER	Wooden wind-oar and twisted whirler and lifter of load.
QUIET	Absent from airmen and easy to horses and got in the grave.
RUDDER	Deflector of flight and flexible fin and pointer of path.
STORM	Night from the north and numbness nearing and hail ahead.
TIME	Expression of alarm and used by the ill and personal space.

UNDER-CARRIAGE	Softener of shock and seat on the soil and easy to injure.
VICTIM	Corpse after crash and carried through country and atonement for aircraft.
WIRELESS	Sender of signal and speaker of sorrow and news from nowhere.
X	Mark upon map and meaning mischief and lover's lingo.
YOUTH	Daydream of devils and dear to the damned and always to us.
ZERO	Love before leaving and touch of terror and time of attack.

Three signs of an airman – practical jokes – nervousness before taking off – rapid healing after injury.

W. H. AUDEN – *The Orators*

In aviation history books the thirties are sometimes referred to as 'the golden years of private flying'. In Britain you could buy a brand-new Moth for six or seven hundred pounds – well within the means of many of the still-prosperous middle class – and get good instruction in how to fly it for not many pounds more. People flew their Moths to the Far East, to Australia, to the Cape, or just to the country for one of those country-house weekends.

Many people have written about learning to fly; as editor of an aviation magazine, I receive a dozen articles a year on the Magic Moment of that First Solo. No one ever did it better than David Garnett, writing in the 1930s in a deceptively simple style. It was still exactly like he described it when I learned to fly on a Tiger Moth in the fifties. Now Moths are highly prized as antiques, and open cockpits and wind-in-the-wires praised as being 'the only true flying'. Maybe it is, at that.

19

FROM: *A Rabbit in the Air*
BY David Garnett
Chatto & Windus, London, 1932

August 31st. 20 mins.
There was a sea mist and a north-westerly wind off land. I got to the aerodrome at 10.45. Clayton told me to get into UH while he started her up. I turned on the petrol.

'Contact,' he shouted. 'Contact,' I yelled back, switching on the impeller magneto, which is the left of the two switches.

'Off.' He swung the propeller round to the compression position. 'Contact.' 'Contact,' I yelled back. He swung and the engine fired. I switched on the other magneto at once, and he climbed in beside me. 'Take her across.' There were some sheep grazing on the far corner.

'How do you look after them?'

'Oh, they keep out of the way if you fly over them. They're well trained.'

'Always turn left on the ground when you turn into wind.' I turned left and looked about to see that we were in wind and that there were no obstacles and no machine landing behind us. Then I opened the throttle and flew off. . . . In a fairly strong wind the machine swings, and one has to keep her straight, and this involved bank and rudder ever so slightly while you're still on the ground.

When we were over the hangars I turned to the Orwell estuary, turned again and flew back. My first glide was just right, but I spoilt it by checking too high up. Clayton put on the engine and took her round, banking sharply about 30 feet off the ground and turning in a small circle. This suddenly gave me complete confidence. The stoical deliberation, the slow precise carefulness in the face of danger, which has been my recent mood, this was all blown out of me by that lovely low racing turn with one wing stretching down to within 15 feet of the turf, and once more I was swept away by the strength and power of the machine – marvellously strong, it would not dawdle about, but fly like a roaring comet, supple and powerful in my hands.

When Clayton shut off the engine and took his hands off the stick, my hand went on controlling it, almost as his had done, and I felt that the machine was full of confidence, brutal certainty and intention, and that it was happy and alive. The invention of dual control has produced a curious heredity in pilots. I learn from Clayton by a 'Laying on of hands', for the feel of his hands on the stick is one of the chief things I learn. So he learnt from another pilot, who in his turn had learnt from one before him. There must be several distinct races of pilots, descended in this way of ordination from the original instructors existing when dual control first became universal. From them all of us are descended.

But Clayton had shut off the throttle and his hands had left the stick as I put her into the glide and eased her up, listening to the 'whick, whick, whick,' of the visibly spinning propeller and watching the earth approaching very fast, with a last final rush it seemed. Then I checked her gradually, held her, got her level, held her while she sank and put the stick back slowly, but alas! just too late – another wheel landing but not at all bad.

The earth approaches very fast . . . that streaming past so marvellously quick that the eye sees only a rushing stream, a turbid liquid of clods and stones melted to porridge to the eye. I had first seen and wondered at that when I was just four years old, and sat with my head lolling over the rail of the governess cart

114

with my eyes fixed on the macadam road, while Shagpat, our New Forest pony, bowled along at a smart trot of nine miles an hour.

The lesson was only twenty minutes. I made several good landings and came back in the afternoon.

NOVEMBER 9TH. 11.15–12.15.

Last night there was not a breath of wind and a hard frost so I looked forward to a lovely clear day, but when I woke I could hear a westerly gale roaring in the elms and the air was mild.

'The wind wouldn't matter if it weren't for these low clouds,' I thought, for there were fingers of hanging mist which raced across the sky, shutting out the occasional glimpses of sun and blue sky.

At the aerodrome I told Honour that I should like to try and find my way across country to Hilton as a holiday from landings, and Marshall gave me a radiant smile and shouted: 'Anywhere you like. Go anywhere you like.'

I took off, brought the machine round about two hundred feet up, caught sight of the railway and began to follow it, looking out for the Huntingdon road on my left. I had never flown before in such bumpy weather or so rough a wind.

'What if I should feel sick?' I asked myself, for the bumps came hard and fast, throwing me up continually against my safety-belt. Holding the machine level was all that I could do.

In clear air it would have been lovely, but sheets of grey vapour whirled down on me and I was lost.

'Keep my elevation by the radiance of the sun and the A.S.I. * constant, and watch the bubble for lateral control,' I said to myself as the cloud swept over us, but when it had passed I was lost and another cloud was sweeping nearer. There was nothing for it but to put the nose down and fly lower. Then I found that railway and followed it. There was still no sign of the Huntingdon road on my left. Presently, however, I saw the water-flooded fields and the river and realized that we must be near Over or Long Stanton, so I bore confidently to the left and suddenly found myself coming over Fenstanton. The Leycester's dovehouse stood up in its green paddock surrounded by trees. I closed the throttle,

---

* Air Speed Indicator.

put the machine into a glide and called out: 'This is Fenstanton,' and then flew off to Hilton, shut the throttle and glided down: 'That is my house.'

It looked very attractive; narrow-backed, ancient, hedged in by elms. The thought came quickly to me that Marshall might not like my flying too low over trees, so I put the engine on and flew over St. John's College farm and round the field which I have decided to use if ever I land a 'plane here. Then I headed for Conington. When we were over the aerodrome. Marshall said: 'Would you like to fly on top of the clouds?' 'Yes!'

I put the nose up into the grey blanket above us and in a moment we were swallowed up and flying blindly. No, not blindly, after all, for there was still a radiance in the whiteness which showed where the sun was and I could judge our angle by that, by keeping the A.S.I. and the Rev. counter both constant, and I watched my bubble. The sun was visible now: a disappearing coin, a silver threepenny bit that came and went like the Cheshire cat, but always leaving its benign smile; a focus of radiance. But better than the sun was the A.S.I.

'If I keep it constant at 65 m.p.h. and the revs. constant at 1800, I must be keeping the nose at the same angle,' I argued.

Meanwhile, the altimeter moved steadily anti-clockwise. At about 1200 feet we came out for a moment and I could see the blue sky above us and we were almost blinded by the brilliant whiteness on all sides. Soon we were right out, clear of the clouds, and I kept her climbing steadily until we reached 4000 feet.

The scene was wonderful, like the scenery for the 'Snow Queen'; a plain of white where nothing stirred, where no living creature would ever set foot, because it was really Heaven. I was in Heaven, out of sight of, and hidden from, the polluting earth.

As far as the eye could see, for hundreds of square miles, stretched the crystalline solid clouds with occasional crevasses between them. The sun blazed with a more than terrestrial glory in an absolutely cloudless sky, and in the distance rose the pointed peaks of Spitzbergen.

'They at all events are real. They can't be clouds.'

The sun wrapped us in gold and striped the fuselage with the shadows of the struts, and keeping it in the same position on my right, I flew with unbounded peace and happiness over this fairy-land, occasionally turning my head to look behind me.

After some little while, lost in ecstatic contemplation, it occurred to me that we must have left Cambridge far behind us, and a few moments later I caught sight of some larger fissures in the ice-floe, through one of which I could see far down, as though at the bottom of the sea, a square field and a long, red-tiled farm building.

I shut the throttle and put the machine into a glide, keeping her air-speed between 60 and 70. The cloud table was woolly now and stretched out wisps of nauseous vapour at me, but I steered away from these clutching fingers to the gulf between. As we descended the air became suddenly bumpy, we pitched and tossed and I took up again the task of fighting to keep the machine level laterally. Below us was a great park with gold and green and bronze trees planted in avenues and clumps.

'Where is the great house?' I wondered, but perhaps it was behind, under the tail, for I saw nothing of it. On my right there was a railway line and I swerved towards it. We had left Heaven and once more were in England. But it was impossible to tell what part. We might be almost anywhere south or east of Cambridge. There had been no means of telling the direction or the strength of the wind above the clouds.

Marshall told me to follow the railway to the right. Flying low along it we came to a tiny station, but there was nothing to tell us what it was. Farther on was a town and a factory, perhaps a brewery, belching smoke and steam from all parts of its roof. Marshall took over the controls and glided down and landed on a field with sheep in it. Landing like this was a great adventure; it made the earth more real. Then we turned and taxied back to the far hedge. Two boys were climbing through a gap in it and a motor car was stopping at the gate.

We were at Bury St. Edmunds.

Marshall climbed out and lifted the tail of the machine round and took off. Then I took over, and picking up the railway line again, flew along it. Beside the railway ran a great straight road with plantations of beeches at intervals. All the way I was fighting with the wind. Presently I saw a racing stable and the slender figures of race-horses in a field. I shut the throttle for a moment to say 'Newmarket.'

117

'Yes, right ahead.'

I turned away from possible churches and flew over the north side of the town over a dozen racing stables. Every paddock seemed to be marked out with a black circle of cinders on which the young horses were exercised. On my right I picked up the great white stand and the great expanse of Newmarket Heath. Near a church on the Newmarket Road I swung away over a fen to avoid disturbing the service and followed the road to the aerodrome. Marshall shut the throttle and began to say something, but I couldn't hear. He slipped off height and landed us. I was tipsy with air when I got out. I was really drugged with oxygen. But my spectacles aren't good enough; I must get goggles with my lenses if I am to fly.

AUGUST 2ND. 10.15–11.0.

There was a north wind and the sky was grey with an overcast sky. Directly I was up, I ran into the mist and began to feel myself lost. After the first landing, I went round two or three times, keeping low all the time and landing well. Then Marshall suggested low flying since conditions were so disagreeable, and we set off. As I got abreast of Girton, Marshall told me to go lower as we were getting lost in clouds, although we were below 300 feet. So I came down low, picked up the Huntingdon Road and flew along it. I held the stick forward, nosing down and towards the road to counteract the drift of the wind. It was marvellous. I was aware, because of the nearness of the earth, of the roaring machine, headlong hurtling, racing over the surface of the earth. Sometimes I brought her really very low and then as I saw trees looming up ahead, lifted her.

At Hilton, I did two turns round the house. I had no eyes for possible members of my family, but only for the elms as I roared off about twice their height.

With the wind behind us on the way back, we were going at the hell of a lick – about 120 miles an hour over the ground, so it was not long before hideous Girton heaved in view. I took her up a bit as I cut across to Chesterton. I shut off and brought her round perfectly to a good landing.

'Time for one more circuit.'

We took off all right, but over the elms the engine missed.

Marshall throttled back, took over and lifted her up so steeply that I spoke my thought: 'What *are* you up to?'

He was gaining height for a conk-out. Then he did a righthand turn, shut the throttle and sang out: 'You've got her.'

I took her in and landed. I was drunk with air. I was wild, and driving home sang and shouted, full of realization that we have found a new freedom – a new Ocean. For thousands of years we have crawled or run on the earth, or paddled across the seas, and all the while there has been this great ocean just over our heads in which at last we sail with joy. The longing for the sea; the call of the sea; one has heard of that, and that was the natural adventure in the past. But now it is a longing for the air, to go up. The air is more marvellous than any sea; it holds more beauty, more joy than any Pacific swell or South Sea lagoon.

Antoine de Saint-Exupery was born into an aristocratic French family. He joined the air force at twenty-one, and flew with them for two years in France and North Africa; then he joined the Air Mail Service, also flying in North Africa. He went to South America and directed the new-born Argentine Air Mail Service, and then returned to Paris and published Vol de Nuit, *which made him a literary celebrity. Attempting to set a record between Paris and Saigon, then French territory, he crashed in the desert and almost died of thirst: from this adventure came his enchanting children's story* Le Petit Prince.

After the fall of France in 1940 he lived briefly and unhappily in New York, before returning to Europe to fly with the Free French forces. He vanished while flying a Lockheed Lightning on a reconnaissance mission to southern France in 1944.

The excerpt that follows is from his South American days in the 1920s.

20

FROM: *Wind, Sand and Stars*
BY   Antoine de Saint-Exupéry
    Heinemann, London, 1939

I had taken off from the field at Trelew and was flying down to Comodoro-Rivadavia, in the Patagonian Argentine. Here the crust of the earth is as dented as an old boiler. The high-pressure regions over the Pacific send the winds past a gap in the Andes into a corridor fifty miles wide through which they rush to the Atlantic in a strangled and accelerated buffeting that scrapes the surface of everything in their path. The sole vegetation visible in this threadbare landscape is a series of oil derricks looking like the after-effects of a forest fire. Towering over the round hills on which the winds have left a residue of stony gravel, there rises a chain of prow-shaped, saw-toothed, razor-edged mountains stripped by the elements to the bare rock.

For three months of the year the speed of these winds at ground level is up to a hundred miles an hour. We who flew the route knew that once we had crossed the marshes of Trelew and had reached the threshold of the zone they swept, we should recognize the winds from afar by a grey-blue tint in the atmosphere at the sight of which we would tighten our belts and shoulder-straps in preparation for what was coming. From then on we had an hour of stiff fighting and of stumbling again and again into invisible ditches of air. This was manual labour, and our muscles felt it pretty much as if we had been carrying a longshoreman's load. But it lasted only an hour. Our machines stood up under it. We had no fear of wings suddenly dropping off. Visibility was generally good, and not a problem. This section of the line was a stint, yes; it was certainly not a drama.

But on this particular day I did not like the colour of the sky.

The sky was blue. Pure blue. Too pure. A hard blue sky that shone over the scraped and barren world while the fleshless vertebrae of the mountain chain flashed in the sunlight. Not a cloud. The blue sky glittered like a new-honed knife. I felt in advance the vague distaste that accompanies the prospect of physical exertion. The purity of the sky upset me. Give me a good black storm in which the enemy is plainly visible. I can measure its extent and prepare myself for its attack. I can get my hands on my adversary. But when you are flying very high in clear weather the shock of a blue storm is as disturbing as if something collapsed that had been holding up your ship in the air. It is the only time when a pilot feels that there is a gulf beneath his ship.

Another thing bothered me. I could see on a level with the mountain peaks not a haze, not a mist, not a sandy fog, but a sort of ash-coloured streamer in the sky. I did not like the look of that scarf of filings scraped off the surface of the earth and borne out to sea by the wind. I tightened my leather harness as far as it would go and steered the ship with one hand while with the other I hung on to one of the struts that ran alongside my seat. I was still flying in remarkably calm air.

Very soon came a slight tremor. As every pilot knows, there are secret little quiverings that foretell your real storm. No rolling, no pitching. No swing to speak of. The flight continues horizontal and rectilinear. But you have felt a warning drum on the wings of your plane, little intermittent rappings scarcely audible and infinitely brief, little cracklings from time to time as if there were traces of gunpowder in the air.

And then everything round me blew up.

Concerning the next couple of minutes I have nothing to say. All that I can find in my memory are a few rudimentary notions, fragments of thoughts, direct observations. I cannot compass them into a dramatic recital because there was no drama. The best I can do is to line them up in a kind of chronological order.

In the first place, I was standing still. Having veered right in order to correct a sudden drift, I saw the landscape freeze abruptly where it was and remain jiggling on the same spot. I was making no headway. My wings had ceased to nibble into the outline of the earth. I could see the earth buckle, pivot – but it stayed put. The plane was skidding as if on a toothless cogwheel.

Meanwhile, I had the absurd feeling that I had exposed myself completely to the enemy. All those peaks, those crests, those teeth that were cutting into the wind and unleashing its gusts in my direction, seemed to me so many guns pointing straight at my defenceless person. I was slow to think, but the thought did come to me that I ought to give up altitude and make for one of the neighbouring valleys where I might take shelter against the mountainside. As a matter of fact, whether I liked it or not I was being helplessly sucked down towards the earth.

Trapped this way in the first breaking waves of a cyclone about which I learned, twenty minutes later, that at sea-level it was blowing at the fantastic rate of one hundred and fifty miles an hour, I certainly had no impression of tragedy. Now, as I write, if I shut my eyes, I forget the plane and the flight and try to express the plain truth about what was happening to me, I find that I felt weighed down, I felt like a porter carrying a slippery load, grabbing one object in a jerky movement that sent another slithering down, so that, overcome by exasperation, the porter is tempted to let the whole load drop. There is a kind of law of the shortest distance to the image, a psychological law by which the event to which one is subjected is visualized in a symbol that represents its swiftest summing up: I was a man who, carrying a pile of plates, had slipped on a waxed floor and let his scaffolding of porcelain crash.

I found myself imprisoned in a valley. My discomfort was not less, it was greater. I grant you that backwash has never killed anybody, that the expression 'flattened out on the ground by backwash' belongs to journalism and not to the language of flyers. How could air possibly pierce the ground? But here I was in a valley at the wheel of a ship that was three-quarters out of my control. Ahead of me a rocky prow swung to left and right, rose suddenly high in the air for a second like a wave over my head, and then plunged down below the horizon.

Horizon? There was no longer a horizon. I was in the wings of a theatre cluttered up with bits of scenery. Vertical, oblique, horizontal, all of plane geometry was awhirl. A hundred transversal valleys were muddled in a jumble of perspectives. Whenever I seemed about to take my bearings a new eruption would swing me round in a circle or send me tumbling wing over wing and I would have to try all over again to get clear of all this rubbish. Two ideas came into my mind. One was a discovery: for the first time I understood the cause of certain accidents in the mountains when no fog was present to explain them. For a single second, in a waltzing landscape like this, the flyer had been unable to distinguish between vertical mountainside and horizontal planes. The other idea was a fixation: The sea is flat: I shall not hook anything out at sea.

I banked – or should I use that word to indicate a vague and stubborn jockeying through the east-west valleys? Still nothing pathetic to report. I was wrestling with disorder, was wearing myself out in a battle with disorder, struggling to keep in the air a gigantic house of cards that kept collapsing despite all I could do. Scarcely the faintest twinge of fear went through me when one of the walls of my prison rose suddenly like a tidal wave over my head. My heart hardly skipped a beat when I was tripped up by one of the whirling eddies of air that the sharp ridge darted into my ship. If I felt anything unmistakably in the haze of confused feelings and notions that came over me each time one of these powder magazines blew up, it was a feeling of respect. I respected that sharp-toothed ridge. I respected the peak. I respected that dome. I respected that transversal valley opening out into my valley and about to toss me God knew how violently as soon as its torrent of wind flowed into the one on which I was being borne along.

What I was struggling against, I discovered, was not the wind

123

but the ridge itself, the crest, the rocky peak. Despite my distance from it, it was the wall of rock I was fighting with. By some trick of invisible prolongation, by the play of a secret set of muscles, this was what was pummelling me. It was against this that I was butting my head. Before me on the right I recognized the peak of Salamanca, a perfect cone which, I knew, dominated the sea. It cheered me to think I was about to escape out to sea. But first I should have to wrestle with the wind off that peak, try to avoid its down-crushing blow. The peak of Salamanca was a giant. I was filled with respect for the peak of Salamanca.

There had been granted me one second of respite. Two seconds. Something was collecting itself into a knot, coiling itself up, growing taut. I sat amazed. I opened astonished eyes. My whole plane seemed to be shivering, spreading outward, swelling up. Horizontal and stationary it was, yet lifted before I knew it fifteen hundred feet straight into the air in a kind of apotheosis. I who for forty minutes had not been able to climb higher than two hundred feet off the ground, was suddenly able to look down on the enemy. The plane quivered as if in boiling water. I could see the wide waters of the ocean. The valley opened out into this ocean, this salvation. And at that very moment, without any warning whatever, half a mile from Salamanca, I was suddenly struck straight in the midriff by a gale off that peak and sent hurtling out to sea.

There I was, throttle wide open, facing the coast. At right angles to the coast and facing it. A lot had happened in a single minute. In the first place, I had not flown out to sea. I had been spat out to sea by a monstrous cough, vomited out of my valley as from the mouth of a howitzer. When, what seemed to me instantly, I banked in order to put myself where I wanted to be in respect of the coast-line, I saw that the coast-line was a mere blur, a characterless strip of blue; and I was five miles out to sea. The mountain range stood up like a crenellated fortress against the pure sky while the cyclone crushed me down to the surface of the waters. How hard that wind was blowing I found out as soon as I tried to climb, as soon as I became conscious of my disastrous mistake: throttle wide open, engines running at my maximum, which was one hundred and fifty miles an hour, my plane hanging sixty feet over the water, I was unable to budge. When a wind like this one attacks a tropical forest it swirls through the branches like a flame, twists them into corkscrews, and uproots giant trees as if they were radishes. Here, bounding off the mountain range, it was levelling out the sea.

Hanging on with all the power in my engines, face to the coast, face to that wind where each gap in the teeth of the range sent forth a stream of air like a long reptile, I felt as if I were clinging to the tip of a monstrous whip that was crackling over the sea.

In this latitude the South American continent is narrow and the Andes are not far from the Atlantic. I was struggling not merely against the crushing winds that blew off the east-coast range, but more likely also against a whole sky blown down upon me off the peaks. Of the Andean chain. For the first time in four years of airline flying I began to worry about the strength of my wings. Also, I was fearful of bumping the sea – not because of the down-currents which, at sea-level, would necessarily provide me with a horizontal air mattress, but because of the helplessly acrobatic positions in which this wind was buffeting me. Each time that I was tossed I became afraid that I might be unable to straighten out. Besides, there was a chance that I should find myself out of fuel and simply drown. I kept expecting the petrol plungers to stop priming, and indeed the plane was so violently shaken up that in the half-filled tanks as well as in the feed pipes the petrol was having trouble coming through and the engines, instead of their steady roar, were giving forth a sort of dot-and-dash series of uncertain explosions.

I hung on, meanwhile, to the controls of my heavy transport plane, my attention monopolized by the physical struggle and my mind occupied by the very simplest thoughts. I was feeling practically nothing as I stared down at the imprint made by the wind on the sea. I saw a series of great white puddles, each perhaps eight hundred yards in extent. They were running towards me at a speed of one hundred and fifty miles an hour where the down-surging windspouts broke against the surface of the sea in a succession of horizontal explosions. The sea was white and it was green – white with the whiteness of crushed sugar and green in puddles the colour of emeralds. In this tumult one wave was indistinguishable from another. Torrents of air were pouring down upon the sea. The winds were sweeping past in giant gusts as when, before the autumn harvests, they blow a great flowing change of colour over a wheatfield. Now and again the water went incongruously transparent between the white pools, and I could see a green and black sea-bottom. And then the great glass of the sea would be shattered anew into a thousand glittering fragments.

It seemed hopeless. In twenty minutes of struggle I had not moved forward a hundred yards. What was more, with flying as hard as it was out here five miles from the coast, I wondered how I could possibly buck the winds along the shore, assuming I was able to fight my way in. I was a perfect target for the enemy there on shore. Fear, however, was out of the question. I was incapable of thinking, I was emptied of everything except the vision of a very simple act. I must straighten out. Straighten out. Straighten out.

One has a pair of hands and they obey. How are one's orders transmitted to one's hands?

I had made a discovery which horrified me: my hands were numb. My hands were dead. They sent me no message. Probably they had been numb a long time and I had not noticed it. The pity was that I had noticed it, had raised the question. That was serious.

Lashed by the wind, the wings of the plane had been dragging and jerking at the cables by which they were controlled from the stick, and the stick in my hands had not ceased jerking for a single second. I had been gripping the stick with all my might for forty minutes, fearful lest the strain snap the cables. So desperate had been my grip that now I could not feel my hands.

What a discovery! My hands were not my own. I looked at them and decided to lift a finger: it obeyed me. I looked away and issued the same order: now I could not feel whether the finger had obeyed or not. No message had reached me. I thought: 'Suppose my hands were to open: how would I know it?' I swung my head round and looked again: my hands were still locked round the wheel. Nevertheless, I was afraid. How can man tell the difference between the sight of a hand opening and the decision to open that hand, when there is no longer an exchange of sensations between the hand and the brain? How can one tell the difference between an

126

image and an act of the will? Better stop thinking of the picture of open hands. Hands live a life of their own. Better not offer them this monstrous temptation. And I began to chant a silly litany which went on uninterruptedly until this flight was over. A single thought. A single image. A single phrase tirelessly chanted over and over again: 'I shut my hands. I shut my hands. I shut my hands.' All of me was condensed into that phrase and for me the white sea, the whirling eddies, the saw-toothed range ceased to exist. There was only 'I shut my hands.' There was no danger, no cyclone, no land unattained. Somewhere there was a pair of rubber hands which, once they let go the wheel, could not possibly come alive in time to recover from the tumbling drop into the sea.

I had no thoughts. I had no feelings, except the feeling of being emptied out. My strength was draining out of me and so was my impulse to go on fighting. The engines continued their dot-and-dash explosions, their little crashing noises that were like the intermittent cracklings of a splitting canvas. Whenever they were silent longer than a second I felt as if a heart had stopped beating. There! that's the end. No, they've started up again.

The thermometer on the wing, I happened to see, stood at twenty below zero, but I was bathed in sweat from head to feet. My face was running with perspiration. What a dance! Later I was to discover that my storage batteries had been jerked out of their steel flanges and hurtled through the roof of the plane. I did not know then, either, that the strips on my wings had come unglued and that certain of my steel cables had been filed down to the last thread. And I continued to feel strength and will oozing out of me. Any minute now I should be overcome by the indifference born of utter weariness and by the mortal yearnings to take my rest.

What can I say about this? Nothing. My shoulders ached. Very painfully. As if I had been carrrying too many sacks too heavy for me. I leaned forward. Through a green transparency I saw sea-bottom so close that I could make out all the details. Then the wind's hand brushed the picture away.

In an hour and twenty minutes I had succeeded in climbing to nine hundred feet. A little to the south – that is, on my left – I could see a long trail on the surface of the sea, a sort of blue stream. I decided to let myself drift as far down as the stream. Here where

I was, facing west, I was as good as motionless, unable either to advance or retreat. If I could reach that blue pathway, which must be lying in the shelter of something not the cyclone, I might be able to move in slowly to the coast. So I let myself drift to the left. I had the feeling, meanwhile, that the wind's violence had perhaps slackened.

It took me an hour to cover the five miles to shore. There in the shelter of a long cliff I was able to finish my journey south. Thereafter I succeeded in keeping enough altitude to fly inland to the field that was my destination. I was able to stay up at nine hundred feet. It was very stormy, but nothing like the cyclone I had come out of. That was over.

On the ground I saw a platoon of soldiers. They had been sent down to watch for me. I landed nearby and we were a whole hour getting the plane into the hangar. I climbed out of the cockpit and walked off. There was nothing to say. I was very sleepy. I kept moving my fingers, but they stayed numb. I could not collect my thoughts enough to decide whether or not I had been afraid. Had I been afraid? I couldn't say. I had witnessed a strange sight. What strange sight? I couldn't say. The sky was blue and the sea was white. I felt I ought to tell someone about it, since I was back from so far away! But I had no grip on what I had been through. 'Imagine a white sea . . . very white . . . whiter still.' You cannot convey things to people by piling up adjectives, by stammering.

You cannot convey anything because there is nothing to convey. My shoulders were aching. My insides felt as if they had been crushed in by a terrible weight. You cannot make drama out of that, or out of the cone-shaped peak Salamanca. That peak was charged like a powder magazine; but if I said so people would laugh. I would myself. I respected the peak of Salamanca. That is my story. And it is not a story.

Harald Penrose was a test pilot for twenty-five years, and for most of them the chief test pilot of Westland Aircraft. He flew 400 different types of aircraft from rotary-engined biplanes to jets; the last he tested was the infamous Wyvern naval turboprop fighter, in which he had seven near escapes, once precariously landing the aircraft with both ailerons deflected full up. He is one of aviation's few 'Renaissance men'; had he not been a test pilot, he might have been an aircraft designer, an ornithologist, or a poet – or a historian, for he is presently engaged in writing a multi-volume history of British aviation.

The chapter which follows is from an earlier book about the birds and skies of Wessex. The flight he describes took place in 1936, in an open-cockpit sailplane he had designed and built himself, a '£30 fantasy of spruce, thin plywood and doped linen'.

21

FROM: *I Flew with the Birds*
BY   Harald Penrose
Country Life/Scribner, New York, 1949

The sunlight made the wings of the white gulls translucent as they soared across the summer blue. While assembling our long-winged sailplanes we watched the birds with interest, for they gave some idea of the extent of the up-currents on which presently we would fly.

At last my own machine was ready. It rested on the turf a hundred yards from the lichened stone wall which rambles along the top of the steep escarpment of Kimmeridge Hill. The scented wind sang in from the sea, two miles away, and leapt over the brow of the hill straight at the sailplane's silver nose, tugging gently at the held-down wings.

When the launching rope had been led from the quick release, around a distant pulley and so to the launching car, I wriggled carefully into the minute cockpit and fastened the safety belt. Soon the ground crew signalled all was ready. A wave of the hand, and the elastic catapult was steadily stretched. I shouted: 'Release!' – and, with a smooth slide, the sailplane shot forward and lifted. Climbing steeply over the wall, into the powerful upsurge of wind deflected by the hill-slope, it freed from the catapult and gracefully made the slightest of dips, as if in salutation to the launching crew and earthbound things. The rushing wind embraced it buoyantly just as the sea holds a boat.

A light pressure on the controls, and the sailplane swung gracefully round until it was flying parallel with the hill ridge. In the strong up-current the machine climbed rapidly for a few moments. As the nose dropped into level flight the loveliness of an enchanting coast began to unfold. Etched against a vast expanse of shining sea the coastline swept ruggedly from St. Aldhelm's Point, changing from shale to smooth curves of chalk as it stretched westward to Portland. Beyond the barrier of that headland the sea was fringed by the parallel curves of the Chesil Beach and the bright strip of the Fleet lake. But the eye swept further – far over the emptiness of West Bay to discover at last a glimpse of white which was Beer Head, the yellow of Sidmouth and Exmouth, and then the dim green of Tor Bay fading into the dark silhouette of Start Point, seventy miles away, below the dim purple shadows of Dartmoor.

Soaring steadily in the hill-wind the sailplane reached the far end of the ridge, canted steeply and returning, swept into a new vista of sunlit fields and moorland and the mirrored surface of the wide waterways of Poole. Like an animated shuttlecock in an infinity of space the sailplane began to sweep steadily to and fro, turning from the unending expanse of the blue Channel seas to the fair prospect of Dorset, and back again to the loveliness of sea.

The air was smooth as ice. With a strange and thrilling sibilance the wind whipped round the gleaming plywood of the cockpit, and curled over the little windscreen, flowing coldly on my face and stinging the eyes. It pressed firmly against the wings, holding the sailplane in cushioned arms, lifting it slowly higher and higher towards the gleaming cumulus that dotted the sky. It seemed that this gentle floating was magic; here at least was affinity with nature and harmony with space. The human shackles had gone – this was the unfettered flight of a bird. . . .

A bird; Better than a bird surely? Higher, faster and quite as agile, I searched the hillside for the soaring gulls we had seen and at once found several, one group being only two hundred feet below me. They gave the impression not of flying but of sliding along the hill-side, their wings held arched and motionless. Slowly, very slowly, the sailplane closed on them – modern wings of polished ply and white fabric competing on equal terms with the gracefully feathered wings of ageless evolution.

Not all birds have the requisite wing shape to enable them to soar in the up-currents generated over the English countryside, but the long-winged gull is one of the greatest exponents. Along cliff-edges in a blustering wind, skimming the steep thrown waves left by a winter gale, or high in the thermal currents of inland regions, gulls are found soaring, flying for miles with barely a wing-beat. Whether they evolved their slender wings to enable them to soar, or soar because they happen to have such wings may be the subject of fruitless debate, but the point of interest is that the aerodynamic characteristics of gulls are exactly matched to the light strength of the up-currents they most often meet. Had the bird a shorter wing, or been of much greater weight for the existing wing area, then soaring would have been impossible except in abnormal circumstances.

As I soared along that fragrant Dorset hill-side, watching the gulls, I wondered again that natural laws enabled a mechanical creation to compete with such airy creatures. Wood, metal, and fabric, used in logical proportion and method, can give strength to withstand the wildest forces of the air and yet weigh, with their human load, no more than the wings can easily sustain by the suction of a moderate speed of air flowing over them. Thus my sailplane could be given wings of such a high ratio of overall span to wing area, yet bearing so moderate a total load, that the up-current required to nullify the machine's natural rate of descent was no greater than that required for the gull. In effect the sailplane dropped a little faster but on a flatter path than the bird. However, where the gull scored was in its ability to reef its wing area, and, by so doing, not only keep just ahead of me, but regulate its forward speed to either breeze or gale.

As I drew level with the birds I glanced at my air speed indicator. The needle was steady at thirty-five miles an hour – a speed giving not my flattest glide, but a slow rate of descent. By going a little slower the sailplane's sinking speed became less than the uprising air, and so the machine began to climb. At 32 m.p.h. I seemed to be keeping station with the birds, but rising at under half a foot a second. I craned over the narrow cockpit-side and watched the gulls intently.

There were three of them in a raggedly extended echelon, and as they flew they eyed sometimes the aerial scene and sometimes each other. Now and again a head would turn towards the wind-humming sailplane above them, and it was a coldly appraising yellow eye that watched. Though each gull was intensely sensitive to everything about it, with reactions set to a hair-trigger, there was no fear of the giant wings above: the sailplane was a flying creature and that seemed sufficient for the bird.

All too soon I reached the ridge-end where I must turn, but the birds swung seaward, and, still holding height, began to traverse the airway above the half-mile of rough meadowland separating hill from shore. I waved farewell.

With nose well down, and a quartering wind to help, the sailplane swept back to the launching point and circled above the watching ground crew. Another gull flew by. I eased the control column a little forward, sensing the elevators pushing against the fluid air – and the sailplane steepened its glide. The wind hummed exultantly as the speed increased to 45 m.p.h. Down and down went the sailplane. In a moment it was skimming the stunted bushes and bracken, slightly rising and falling in the rougher air just above the ground. Smoothly sliding, as though on ice, it raced to the downwind end of the hill ridge. As Swyre Head came into view I pulled up in a climbing turn almost on to the tail of a soaring herring gull.

Not more than forty feet ahead, and a little higher than the glider, the gull sailed nonchalantly on. It glanced under a wing, then gave three lazy flaps. The impetus caused it to rise a few feet. I followed. The bird swept up a couple of yards. The sailplane closed a foot or two. With half a dozen wing beats the gull made good the difference, climbed a little higher – and settled to its soaring.

I turned the sailplane further from the hill face, so that our flight paths were parallel, but some fifty feet apart. Slowly the glider began to catch the bird. At 37 m.p.h. I was barely holding height. My speed dropped to 34. Much slower now, the distance between us lessened. Soon the bird was not more than ten feet ahead but on a course someway beyond my wing tip. It gave a few

132

easy flaps and soared again. I rose on a gust. The bird climbed too, and then flapped for a moment to put itself well above me. From twenty feet higher it stared at me as we floated in formation past the ground crew, past the wind-sock, past the last of the bracken, the beginning of the pasture, and then swept steadily around the long crescent of the hill.

'Cheat!' I called to the bird. 'Stop flapping and I'll beat you at your own game.' But the gull went sailing on, with on occasional wing beat that kept him master of the situation until presently we reached the end of my circuit. Like those others, the gull went soaring on towards the sea.

The next time I came to that point I also turned seawards. As though sailing a placid mill-pool the glider went soaring away from the hill, holding its height just like the gull. Away and away from the up-currents of the hill it went, held up on a great mass of sun warmed air rising from the fields between the cliffs and the hills. Cautiously exploring I began to find that in this area I could soar wherever I willed.

Presently I steered beyond the cliffs, nosing over the sea until the coastline lay half a mile behind and there seemed only blue around to the world's end, with the harsh rasp of the waves overpowering the gentle sound of the sailplane's flight.

In sudden apprehension I swept away from the endless water, speeding in a great curve back to the safety of the sun warmed land. Kimmeridge Hill seemed far away – impossible to reach. But the gentle thermal still held the glider's wings, and the sea-breeze was now a tail-wind helping the race for home. Fields, trees, hedges, cows, flashed under the nose and were lost behind from sight. The ground began to rise. Suddenly the sailplane nosed into the strong up-current of the hill and rocketed above the summit, four hundred feet in half a minute.

Though I had intended to land, such a gift of height was too valuable to throw away at once, so four times more I soared along the hill, gazing at the loveliness of Wessex spreading wider and wider, until it was hidden in the purple shadows of the far horizons.

Reluctantly I turned the sailplane into the descending air behind the hill-top. On a smooth, heather-covered area my friends were ready to grasp the wings on landing. They stood motionless, their upturned faces white against the brown ridge.

Turning in smooth dropping curves, to left and right, the sailplane slid down its invisible air-slope towards the heather. One hundred feet up, fifty. Yes! going to land just by the crew! Twenty feet, ten, five, one – and with a fading sigh the sailplane lightly touched the ground and stopped.

For a moment the heather and long bracken, the distant shrubs, the tumbled stone wall marking the hill edge in front, were invested with an air of unreality. I looked up at the sky. White-winged, tranquil, effortless, a gull sailed by. Suddenly I was landbound again, and I stared at the bird as though it were a creature of magic and only the earth was real.

Airline pilots today earn as much as $80,000 (£46,000) a year and get their life insurance at the same rates as everyone else. Their life was nothing like so lush when Ernest Gann was a co-pilot on DC-2s in the 1930s. 'There are two kinds of airplanes,' his captain explained, 'those you fly, and those that fly you. With a DC-2 you must have a distinct understanding at the very start as to who is boss . . . You will learn to love this airplane; and you will also learn to hate it.'

The apprenticeship whereby co-pilots learned their trade from their captain through osmosis, admonition and example was, said Gann, 'uneasy labour for both men, (which) occasionally took on the character of an unrehearsed wrestling match.' Gann was usually assigned to a Captain Ross, 'a fierce and dedicated tyrant. His rebukes echoed even in my sleep. He never relented in his instruction which had the quality of ceaseless pounding, so that frequently at the end of a flight my brain seemed to hang limp between my ears, twisted and bruised.' It was in Ross's crew that Ernest Gann first experienced the savagery of a squall line thunderstorm.

22

FROM: *Fate is the Hunter*
BY Ernest K. Gann
Hodder & Stoughton, London, 1961

It is also the commencement of the thunderstorm season, and here, in the region between the eastern sea coast and Cleveland, such opponents are only surpassed in ferocity by those truly evil monsters whose lair is farther to the south, between Washington and Memphis. Others of like pugnacity lurk about the mid-western states, but a thunderstorm is a thunderstorm no matter where encountered, and all of them have the disposition of a Caligula.

Any pilot of sound mind avoids flying through a thunderstorm, if he can possibly find a tunnel of escape. Unfortunately, there are times when the fray is unavoidable, and this is particularly so on AM-21 where climbing to a very high altitude is usually out of the question. On certain midsummer afternoons and nights, the route becomes a jungle roaring with resentment at all intruders.

And now, an airport long abandoned for inadequacy has been re-established along the route. This is Albany, and between it and Newark is to be found an area of thunderstorm activity unmatched anywhere on earth, including the ill-famed west coast of Africa. For here the route follows the Hudson River southwards, passing over the Catskill Mountains where the legendary ghosts of Dutchmen bowl with such abandon. In this region a special combination of terrain, summer heat and humidity, serves to produce spectacular monuments of clouds. Pilots who would much prefer admiring this grandeur from afar, frequently refer to such formations as sons-of-bitches. The pilots know they are not be trifled with for they can, on occasion, make a little boy trembling beneath his bed of the boldest man.

It is here, with Ross, that I first witness the anger of God in new dimensions.

We are southbound for Newark at eight thousand feet. The sun is gone, but enough light lingers in the sky to illuminate the purple valley below which embraces the Hudson. In the east the sky is fast losing the violet hue of twilight and ahead, lying at right angles to our course, is visible confirmation of the weather report received in Albany. From horizon to horizon the southern sky appears supported by a series of castles, battlements and walls. Through the various indentations a few jagged gashes of the lower sky may still be seen. These are a sickly yellow, and above the highest escarpments the sky fades away rapidly towards the zenith until it becomes a pale blue tinged with black. The very summits of the highest cloud towers, of which there are two most prominent, still catch the escape of the sun and are golden.

I have completed the necessary entries in the log-book, and in spite of the insane crackling in my headphones believe that I can recognize the easy voice of McGuire reporting the position of his plane over South Bend, westbound for Chicago. I am intrigued with the idea that the sun is doubtless still in his eyes and that he might soon be sipping a beer in a place near the Chicago Airport called the 'Snake-pit'.

I sit far back in my seat, my right foot braced comfortably against the instrument panel, listening to the steady thrumming of the engines, content to reflect that I have at least come a long way since the days of barnstorming. Not so long ago, in a rock-fenced field near by, a young man named Blauvelt stepped away from a sputtering biplane and first sent me into the sky alone.

It is pleasant to think that now my ninety-day probation is safely past and as a consequence I might reasonably invest in a pair of jodhpur boots like Ross's, and so keep my ankles warm during the coming winter. With concentration I might also acquire that certain swagger which must go with the wearing of such boots. I will break them in very carefully since they are extremely expensive for a co-pilot's purse, and I will try to keep them shined as well as Ross's.

Staring at the gigantic panorama ahead, my mood attempts to match it and soon becomes so expansive as to ignore all reality. I fancy that I can handle a DC-2 in the air and on the ground as well as, perhaps better than, most men. I am being paid for moments like this as well as for the lesser times, and though my station is humble no man commands me for whom I fail to hold the deepest respect. This in itself, is certainly a most unusual and agreeable situation.

As we draw closer to the array of clouds ahead, and even the minor patches of sky are obscured by over-reaching claws of vapour, my musings take a truly heady turn. There is a new rumour that the line is about to purchase ten more aeroplanes. Applying an airline rule-of-thumb, ten new aeroplanes will require three crews each. That will be thirty new Captains then, promoted from co-pilot. There are approximately one hundred co-pilots with a number lower than my own. Therefore, assuming the line continues to expand at the rate of ten aeroplanes per year, why shouldn't I make Captain in about three years? What a prospect! Magnanimous in my dreams, I resolve to treat my co-pilot with the utmost consideration, thus remembering my own uneven days of servitude.

These rich speculations are suddenly interrupted by a new and heavy surging from the engines. Ross has shoved the propeller controls into low pitch and is advancing the throttles. I see that he has put on his gloves and we are climbing to the best of ability of the DC-2. He flips the switch which illuminates the seat-belt sign and nods his head towards the cabin door.

'Go back. Tell the stewardess to make sure the passengers are

137

really strapped in and haven't just got the belt laying across their legs.'

There are only nine passengers. I move down the line of their inquiring eyes in a terrible parade of self-consciousness. Like most pilots I have already learned to shy away from contact with passengers; their queries being often so penetrating that a straightforward answer is only reasonable. The results are sometimes more embarrassing than satisfying.

When my message is delivered I start back to the cockpit. A young woman sits near the door. She is peacefully breast-feeding a baby and like most men the discovery somehow astonishes me. I look instantly away. The scene makes me strangely uncomfortable, yet as I resume my seat beside Ross, I cannot put it out of my mind. The young mother was far from beautiful, and still, in her absolute trust, she *was* beautiful. And the trust was in Ross, and to a degree, myself. I discover that it is the kind of faith I would just as soon not think too much about.

Now, looking at Ross, I see him in a new light. The difference is in his face which has taken on a sternness I have never before observed. His entire body seems to gather itself as he squirms down slightly in his seat. He hunches his broad shoulders, holding and then releasing them, a flex of body much like a boxer stepping from his corner. He lights a cigarette, takes a few puffs, then smashes it into the ash-tray at his side.

His eyes are also new to me. Because he is constantly moving his head searching all about the sky, it is difficult to appraise his thoughts, but his eyes are unusually alert and seeking, as if somewhere in the gathering gloom ahead he has detected a physical threat to his well-being. Ross does not bear the spoke-like glare wrinkles about his eyes common to so many pilots, but now the flesh there becomes pinched in half mischievous smile.

'Call Newark. Get us a clearance for twelve thousand. Ask for their weather while you're at it.'

I comply, breathing between my requests more frequently than normal because we are already at ten thousand feet, an altitude for which my AM-21 service has ill-accustomed me.

The voice from Newark is barely intelligible through the barrage of static in my headphones. It is gibberish orated from the bottom of a barrel and I can only be certain that the clearance for

138

altitude is granted. Fragments of the Newark weather gurgle through, sounding like the saliva-laden mouthings of an imbecile. 'Thunderstorms over . . . heavy rain . . . visi . . . lightning to the . . . wind. . . .'

Nothing more.

Ross nods his head. He has apparently understood where I could not. And he has more important matters on his mind. His entire attention is devoted to the slowly floating scenery beyond the windshield and his prediction is simple enough.

'I think we are going to take a pasting.'

We are still some three or four miles from the immense barricade, yet even at this distance the appalling energy contained within the dominant sections of cloud is all too apparent. There is nothing static about these pillars so hugely piled one on top of the other. They are in constant motion, not horizontally as wind might activate them, but appearing to roll about a series of central axes, each pillar enlarging and twisting upon itself.

As we approach closer the vaporous appearance proves a complete deception and each rolling protuberance takes on a solid look like tremendous pus-laden pustules in the moment before eruption. Yet the air is still remarkably smooth, as if the barrier were spreading a carpet of invitation.

Quite suddenly, when we are barely a mile from the nearest tower, all sense of established proportion is reduced to the truth. The effect on my ego is stunning. For it is a value of normal flight that human beings carry aloft their usual visionary sense of relation between themselves and those extraneous objects familiar to their eyes. In an aeroplane cockpit, whether it be on the ground or in the air, the instrument panel remains the same size, as does the throttle, the controls, the windows, and the wings. Likewise, the accompanying humans hold their stature. It is a cramped world in itself, but everything within it is quite standard and therefore bearable. All else beyond the windshield is so far away as to appear itself diminutive, rather than the other way around.

Now, approaching this thunderstorm front which is neither larger nor smaller than any other common to the season, our comfortable world is lost and no endeavour can bring it back

139

again until we are either in the clear or cannot see at all. We are suddenly so puny we belong on glass, beneath a powerful microscope. The sensation is shocking, the escape of conceit from our being, instantaneous. How can such infinitesimal creatures presume to trouble the heavens with our mewling hopes and complaints? For here, alongside mightiness, we are nothing.

Our altitude of more than two miles above the earth is less than half that of the most prominent thunderheads. The phalanx forms a solid precipice which tumbles straight down from the edge of our wing-tip, grey-black and green in the last light. Great blossoming fists of dirty white churn against each other all the way to its gloomy foundations. Inside the darker areas there are frequent explosions of light, marked simultaneously by savage crashes in our earphones.

Ross sighs heavily and regards the clouds with deep suspicion. He shakes his head sadly in the manner of a man affronted by a sidewalk bully. Here is work, unpleasant work, perhaps even dangerous if the bully gets out of hand, but the struggle is unavoidable.

The more realistic line pilots freely admit that most of the time they are overpaid. It takes a while to discover this life is somewhat short of paradise. For there is, as in all pursuits of mankind, no true ideal or even near perfection. In airline flying the hitch always comes, and it follows an apparently unshakeable cycle of once a year. A pilot may earn his full pay for that year in less than two minutes. At the time of the incident he would gladly return the entire amount for the privilege of being elsewhere.

Now it appears that Ross may more than earn his wages for this August evening.

He banks the ship steeply and for a time we fly along parallel to the unbroken cliff. And though we are proceeding at normal cruising speed we seem to hang dreamlike and nearly still in the air.

At last a gaping crevasse appears. It is guarded on both sides by vaporous gargoyles projecting from Byzantine towers. Though tortuous, it seems an offer of good entrance, yet Ross dismisses it with a glance.

He carries on perhaps a mile until a second chasm appears which is darker, though not so confined. He switches on the cockpit lights, turns them up to full bright, and lowers his seat all the way to the floor. I wonder if he knows this is my first encounter with a thunderstorm. He points to the position of his seat.

'I recommend you do the same.'

'Why?'

'You will see.'

So hunched, our faces below the bottom edge of the windshield and thus somewhat protected, we enter the division.

I am not afraid, but reason that it is perhaps just as well I am not yet a Captain.

For a few minutes our flight remains quite peaceful although there are minor bumps which are scattered and uneven like rocks strewn before the entrance of a cavern. Looking upward from my side window I can still see the formations which compose this entrance into the mass, and by twisting against my seat-belt I can observe a patch of clear sky some ten thousand feet above all. The fragment is now the colour of slate.

Ross's attention remains entirely within the cockpit. He is flying on instruments and has taken up an approximate course for Newark. He is not tense, but very alert. He seems still to be waiting. I wish Ross would stop exploring the sky and get this aeroplane on the ground in Newark. I need a cool shower to draw away the heat and work of the day, and I am also very hungry.

I almost resent it when Ross suddenly reaches for the throttles and pulls them back. Our speed slows to one hundred and forty miles an hour.

Light rain hisses along the windshield, sounding for a moment like a disturbed nest of snakes. Ross calls for heat to the engine carburettors. I pull back the two red-knobbed levers on the side of the control pedestal and adjust them so that the temperature gauges read properly.

I am so engaged when our small world goes insane. Some preposterous genie turns a fire hose full on the windshield. We are suddenly not in an aeroplane, but a submarine – one that leaks very badly. Water spews through the nose, the windshield scarfing, my wide window, the roof. It dribbles down the instrument panel, sopping our pants. But it is the sound that chokes off my half-spoken curse of protest. It is a roaring of water, an angry, bewildering, sense-shattering cascade that completely obliterates every other noise. It is inconceivable that the engines can swallow so much liquid and continue to function. Or that we can maintain flight through the depths of an ocean.

I glance at Ross. He has both hands firmly on the control wheel, but he still seems to be waiting. I have a new and strangely sour taste in my mouth. I am wondering about it when the bottom falls out of everything.

We seem to smash against a solid obstruction. I am instantly weightless, jerked hard against my seat-belt. The instrument panel shivers so beneath the shock that for a moment not a dial is readable. My confused eyes seek some reassurance in the turmoil. Ross, please! Bring back stability and reason!

When the instruments settle I am appalled at the altimeter and rate of climb. In spite of Ross's strenuous efforts at the controls, we are going down fifteen hundred feet per minute! The altimeter continues to unwind. Ross! This is not good. I am not afraid, of course, but certain of my glands are misbehaving. And the noise. My brain is pressing down on my eyes. It is difficult to think, see, or hear anything that is reliable.

Ross calls for more power and shoves the propeller controls to full low pitch. There is no familiar howl from the engines. This puny sound cannot predominate the roar of water. Only the instruments show they have responded.

Ross wrestles with the control wheel. He seems unperturbed though defiant. He rolls the stabiliser wheel back a few turns, trying to ease the ship into a climbing attitude. According to the power we should be ascending at least six hundred feet per minute.

We do not rise an inch. Our descent continues as if the ship were actually in a dive. We are within the grasp of power far more formidable than the Wright engine company can ever produce.

We continue down sickeningly, to nearly ten thousand feet.

Then a second collision occurs. Once more the instrument panel dances in its rubber mountings. The dials blur, then shake themselves back into readability. To my amazement we are now going up at fifteen hundred feet per minute. Reversing itself, the altimeter winds back and passes eleven thousand in spite of Ross shoving the nose down hard. He raises one hand and makes a deliberate gesture towards the cockpit floor. He yells above the pandemonium.

'Gear down!'

Has he completely lost his wits! You can't, not even you, Ross, land in mid-air. He repeats the command. Since he obviously means it, I oblige.

Now he literally stands the ship on its nose. We are gear-down in a steep dive. It becomes a ludicrous flight attitude when the instruments relay the very definite information that regardless of all efforts, we are still going up.

I soon appreciate the fact that were it not for the gear-down and our consequent steeper angle of attack, we would be going up much faster.

As we pass through twelve thousand feet again my uneasy nerves are jolted by a new display. The roar of rain diminishes as suddenly as it came and I am momentarily blinded by an explosion of white fire which seems to occur directly within the cockpit. If Ross had not turned the lights up full bright, the blinding would last much longer. And we receive some protection from the lowered seats. I silently thank him for his foresight and store the trick in my memory.

The lightning continues in a series of flashings each accompanied by a blasting cannonade. I discover very quickly that thunder at the source is not in the least like thunder heard on the ground. There is nothing to provide an echo, hence the detonation is utterly flat in tone. Every salvo pierces straight to the soul. It is a hellish tympany and you wish you were deaf.

I am cold and there is a strange ache in my belly when we approach thirteen thousand feet with the ship still pointed down. Our lateral gyrations now become extremely violent. Tossed beyond its limits, the artificial horizon instrument 'tumbles', and becomes therefore useless. Ross is forced to fly on only turn indicator and air speed, an arduous commitment in such rough air. I prefer not to look back at the wings, even if I could see them. Some things are better unknown.

Our ascent slows and I breathe a sigh of relief which is cut short by a head-on collision with an express train. All previous noise is now insignificant. We are in hail.

I see Ross's lips move in malediction, but I cannot hear his

143

voice, A thousand machine-guns are directed upon us and we are in a tin can.

The hail is of short duration, perhaps one minute, or two at the most. The stones are not big enough to dent the ship's skin, which is by no means an unusual development, but they are sufficiently concentrated to remove the paint from the nose.

The unpleasantness continues for ten minutes. Then, as if the elements had wearied of playing with us, all action ceases. The air smooths as suddenly as it revolted. We are in heavy cloud, but there are only occasional short spasms of rain. We glide down to proper altitude easily, in seeming silence. After so much noise the engines merely purr.

In my headphones there is still an uncomfortable crashing, but the Newark beam is easily detectable. A voice from our company radio is superimposed upon its reassuring whine, and the words spoken are quite intelligible. The voice is reciting the Newark weather to a plane in the vicinity. 'Ceiling four hundred feet light rain, visibility two miles . . . altimeter twenty-nine ninety.'

We set out altimeters accordingly and Ross calls for a clearance to six thousand feet. When it is received he waves the controls over to me. This is a mild shock for it is his leg and therefore the instrument approach which the Newark weather demands should be his. I had thought to coast the rest of the way.

'We will now learn what it's like to make a real approach without an artificial horizon' – he points at the offending instrument which swims uselessly behind its glass face – '. . . and a few other things which may some day be worth our knowing.'

I cannot understand Ross. He is obliged to give nothing away – much less an instrument approach which could be difficult. He is not paid to be a mentor and can, in the doing, only risk his peace of mind.

'Now concentrate. Forget I'm here.'

Forgetting Ross is like a slave forgetting the galley master, in spite of the fact that he tips back his head, closes his eyes, and seems to be napping. Yet in no other way can he appear to set me on my own.

A radio range is simply a broadcasting station which transmits a monotonous though valuable programme. The signals form four spokes, each narrowing and intersecting at the hub like an ancient

144

wheel. The spoke selected depends, of course, upon the direction from which the plane approaches the station. I pick up and begin to bracket the north spoke of the Newark range.

Basking in Ross's confidence, I move the controls gently. I am only too aware that his closed eyes are a sham. He is also listening, and like a temporarily relaxed conductor, waiting for a sour note.

When I have at length settled down to a steady course, Ross opens his eyes and presses a button on the instrument panel three times. In a moment the stewardess appears in the shadows behind us. He asks her about the passengers and how they fared through the storm.

'One man was sick. They are laughing about it now.'

Without turning from my work, I ask about the baby. I don't really care, but somehow I must know if it bawled.

'No. It slept all the way through. Anyone you know?'

'No. I was just curious.'

But I brooded on this for several minutes and could not decide why it seemed so wonderful.

We pass over the cone of silence at Newark with very little meandering, once I have pinned down the leg. The initial descent is pleasing too, steady and nicely timed and I can reasonably hope the rest of the performance will go as smoothly.

Then as we start the turn for the final descent, which always is the most complicated and demanding in accuracy, Ross takes a box of matches from his pocket and lights them one after another just under my nose. I gasp a protest. I am heavily engaged in trying to hold course and altitude exactly according to the book. This is the real thing. It counts.

'What the hell are you doing?'

I am bewildered. If I were not so extremely busy I would brush the flame away. It is difficult to see the instruments beyond the flame and Ross holds it just close enough to make breathing difficult.

I blow out the match. Ross at once lights another. I am fifty feet too low, the compass is swinging in a direction it should not, and my speed is falling off.

'Steady . . .'

Ross's voice is calm and without malice or mischief. Then what in God's name is he up to? The performance, on which I was just about to congratulate myself, is rapidly going to pieces.

I fight to keep things in order, not because we are in the slightest danger at this altitude, but only because Ross has deliberately ruined what might have been a technically perfect approach. For this I cannot forgive him.

As one match after another flares before my eyes I become infuriated with Ross. He is a sadist; sick with weird complexities. He is afraid I *will* do a good job. To hell with him! I will keep all as it should be regardless of his jealous interference.

Sweating profusely, inwardly cursing Ross's twisted sense of humour, I resolve to fly this ship safely and surely to earth in spite of any harassment. I force myself to ignore Ross's match – see beyond it to the instruments.

As we turn in for the final descent I shove the propeller controls to full low pitch. We are exactly at required altitude, the speed is right, and also the course.

Ross shakes out his match and sits back in his seat. I glance at him, my resentment doubling when I discover him smiling. We will have this out on the ground!

In less than a minute, at six hundred feet, the faint glow in the clouds becomes an iridescent bloom. I hold the descent. Tatters appear in the cloud base, then the runway lights and finally the guiding ladder of red neon tubes dead over the nose. I call for full flaps, chop off the power, and we swoop down through light rain until the wheels brush the cinders. I believe that even McCabe would class the landing as better than an arrival, although so short a time ago I strained his back on almost the identical spot.

As for Ross, he can take his comic-opera cap and fly in other directions. I intend to ask for a transfer.

When the engines are stopped I complete the log-book in wounded silence. Ross leaves his seat and puts on his coat. It is raining harder outside. Maybe his ridiculous cap will shrink to the size of his brain.

I snap the log-book shut and am about to stand up when I feel his heavy hand on my shoulder. My grip on the metal log-book tightens. If he tries one of his playful swings . . .!

But his voice is surprisingly tired and so is his smile.

'Anyone can do the job when things are going right. In this business we play for keeps.'

When he has left the cockpit I remain in my seat listening to the rain peckle on the aluminium above my head. The matches. Why would he light matches? He could more easily have created other distractions if that had been his only intent.

I walked slowly through the rain to the operations office, not really caring if my uniform was further soaked. I decided against asking for a transfer. Ross, in his peculiar way, was making a line pilot of me. And I supposed it was a good way.

Nearly four years would pass before I would again see Ross's matches flaming before me. Then, even though distracted by the wild drumming of my heart, I would know their incalculable worth.

*John Gillespie Magee was an American serving with the RCAF; he
was just nineteen when his Spitfire rammed another aircraft in cloud
in 1941. Among his personal effects was an envelope with this simple
sonnet scribbled on the back.*

23
Dancing the Skies
BY   John Gillespie Magee

Oh! I have slipped the surly bonds of earth
   And danced the skies on laughter-silvered wings;
Sunward I've climbed, and joined the tumbling mirth
   Of sun-split clouds – and done a hundred things
You have not dreamed of – wheeled and soared and swung
   Hung in the sunlit silence. Hov'ring there
I've chased the shouting wind along, and flung
   My eager craft through footless halls of air.

Up, up the long, delirious, burning blue
   I've topped the wind-swept heights with easy grace
Where never lark, nor even eagle flew –
   And while with silent, lifting mind I've trod
The high, untrespassed sanctity of space,
   Put out my hand and touched the face of God.

*Richard Hillary had belonged to that generation of Oxford
undergraduates who had earlier voted after a debate 'that this house
would not fight for king or country'. But fight they did when the time
came. As a fighter pilot Hillary was shot down into the North Sea,
grievously injured with dreadful disfiguring burns. He had always
wanted to be a writer. He recovered sufficiently to return to
operations, but was killed before he'd had time to write more than
this one book. The excerpt describes his first agonies after being
burned; only the first, be it noted.*

24

FROM: *The Last Enemy*
   BY   Richard Hillary
        Macmillan, London, 1942

I was falling. Falling slowly through a dark pit. I was
dead. My body, headless, circles in front of me. I saw it with my
mind, my mind that was the redness in front of the eye, the dull
scream in the ear, the grinning of the mouth, the skin crawling on
the skull. It was death and resurrection. Terror, moving with me,
touched my cheek with hers and I felt the flesh wince. Faster,
faster. . . . I was hot now, hot, again one with my body, on fire and
screaming soundlessly. Dear God, no! No! Not that, not again.
The sickly smell of death was in my nostrils and a confused roar of
sound. Then all was quiet. I was back.

Someone was holding my arms.

'Quiet now. There's a good boy. You're going to be all right.
You've been very ill and you mustn't talk.'

I tried to reach up my hand but could not.

'Is that you, nurse? What have they done to me?'

'Well, they've put something on your face and hands to stop
them hurting and you won't be able to see for a little while. But
you mustn't talk: you're not strong enough yet.'

Gradually I realized what had happened. My face and hands
had been scrubbed and then sprayed with tannic acid. The acid
had formed into a hard black cement. My eyes alone had received
different treatment: they were coated with a thick layer of gentian
violet. My arms were propped up in front of me, the fingers

extended like witches' claws, and my body was hung loosely on straps just clear of the bed.

I can recollect no moments of acute agony in the four days which I spent in that hospital; only a great sea of pain in which I floated almost with comfort. Every three hours I was injected with morphia, so while imagining myself quite coherent, I was for the most part in a semi-stupor. The memory of it has remained a confused blur.

Two days without eating, and then periodic doses of liquid food taken through a tube. An appalling thirst, and hundreds of bottles of ginger beer. Being blind, and not really feeling strong enough to care. Imagining myself back in my plane, unable to get out, and waking to find myself shouting and bathed in sweat. My parents coming down to see me and their wonderful self-control.

They arrived in the late afternoon of my second day in bed, having with admirable restraint done nothing the first day. On the morning of the crash my mother had been on her way to the Red Cross, when she felt a premonition that she must go home. She told the taxi-driver to turn about and arrived at the flat to hear the telephone ringing. It was our Squadron Adjutant, trying to reach my father. Embarrassed by finding himself talking to my mother, he started in on a glamorized history of my exploits in the air and was bewildered by my mother cutting him short to ask where I was. He managed somehow after about five minutes of incoherent stuttering to get over his news.

They arrived in the afternoon and were met by Matron. Outside my ward a twittery nurse explained that they must not expect to find me looking quite normal, and they were ushered in. The room was in darkness; I just a dim shape in one corner. Then the blinds were shot up, all the lights switched on, and there I was. As my mother remarked later, the performance lacked only the rolling of drums and a spotlight. For the sake of decorum my face had been covered with white gauze, with a slit in the middle through which protruded my lips.

We spoke little, my only coherent remark being that I had no wish to go on living if I were to look like Alice. Alice was a large country girl who had once been our maid. As a child she had been burned and disfigured by a Primus stove. I was not aware that she had made any impression on me, but now I was unable to get her out of my mind. It was not so much her looks as her smell I had continually in my nostrils and which I couldn't dissociate from the disfigurement.

They sat quietly and listened to me rambling for an hour. Then it was time for my dressings and they took their leave.

The smell of ether. Matron once doing my dressing with three orderlies holding my arms; a nurse weeping quietly at the head of the bed, and no remembered sign of a doctor. A visit from the lifeboat crew that had picked me up, and a terrible longing to make sense when talking to them. Their inarticulate sympathy and assurance of quick recovery. Their discovery that an ancestor of mine had founded the lifeboats, and my pompous and unsolicited promise of a subscription. The expectation of an American ambulance to drive me up to the Masonic Hospital (for Margate was used only as a clearing station). Believing that I was already in it and on my way, and waking to the disappointment that I had not been moved. A dream that I was fighting to open my eyes and could not: waking in a sweat to realize it was a dream and then finding it to be true. A sensation of time slowing down, of words and actions, all in slow motion. Sweat, pain, smells, cheering messages from the Squadron, and an overriding apathy.

Finally I was moved. The ambulance appeared with a cargo of two somewhat nervous A.T.S. women who were to drive me to London and, with my nurse in attendance, and wrapped in an old grandmother's shawl, I was carried aboard and we were off. For the first few miles I felt quite well, dictated letters to my nurse, drank bottle after bottle of ginger beer, and gossiped with the drivers. They described the countryside for me, told me they were new to the job, expressed satisfaction at having me for a consignment, asked me if I felt fine. Yes, I said, I felt fine; asked my nurse if the drivers were pretty, heard her answer yes, heard them simpering, and we were all very matey. But after about half an hour my arms began to throb from the rhythmical jolting of the road. I stopped dictating, drank no more ginger beer, and didn't care whether they were pretty or not. then they lost their way. Wasn't it awful and shouldn't they stop and ask? No, they certainly shouldn't: they could call out the names of the streets and I would tell them where to go. By the time we arrived at

Ravenscourt Park I was pretty much all-in. I was carried into the hospital and once again felt the warm September sun burning my face. I was put in a private ward and had the impression of a hundred excited ants buzzing around me. My nurse said good-bye and started to sob. For no earthly reason I found myself in tears. It had been a lousy hospital, I had never seen the nurse anyway, and I was now in very good hands; but I suppose I was in a fairly exhausted state. So there we all were, snivelling about the place and getting nowhere. Then the charge nurse came up and took my arm and asked me what my name was.

'Dick,' I said.

'Ah,' she said brightly. 'We must call you Richard the Lion Heart.'

The quickest-witted, most thoughtful fighter pilots were usually the most successful, with the highest scores of kills and the best chance of surviving the war. Some of these later wrote of their experiences; and a couple, Pierre Clostermann and Johnnie Johnson, went on to write further of the general history of air fighting. Clostermann was a Frenchman who joined the Free French forces in Britain after the fall of France, and flew the extraordinary number of 420 missions against the Germans, beating enormous odds against his surviving the war to go on to become the head of the Cessna lightplane factory in France afterwards.

Here he describes a little-known aspect of the Japanese raid on Pearl Harbor: that the approaching aerial armada was shown on a pioneer radar set which the British had lent the US Navy; its terrified operator could not find anyone in command to take his report seriously.

25

FROM: *Flames in the Sky*
BY Pierre Clostermann
Chatto & Windus, London, 1952

7th December 1941

It was Sunday morning – the first Sunday after payday for the G.I.s and sailors of the base. All night the dance halls and bars of Pearl Harbor had been turning people away. The Fleet had come in the day before – apart from the aircraft-carriers – and Hawaii had been the scene of the usual junketings on the first Saturday in the month.

There were few people still up at 3.45 a.m., apart from the poker addicts and those wending their way back to their Mess.

In a tent on a hillside overlooking the misty sea an alarm-clock went off. Technical Corporal 3rd Class Joe Lockard and Private George Elliot climbed grumbling out of their damp blankets and after a quick wash went to their post in the chilly morning air.

By their tent stood a large square steel trailer with narrow slatted windows. On the roof, covered with a tarpaulin which the men carefully removed, was a large parabolic aerial shaped like an electric bowl fire. It was one of the three SCR 260-B experimental radar sets which had arrived from England. Nobody had much faith in this queer British contraption. Joe Lockard, an amateur radio fan in private life, was the only one – or at any rate one of the few – for whom the apparatus was now practically an open book.

The radar sets had arrived in July and had been on the go continuously from 10th November to 3rd December whenever an 'alarm' warning had been given. Handled as they were by inexpert crews, they had begun to go wrong and there were now only three spare cathode-ray tubes. H.Q. had decided to use them only from one hour before dawn till one hour after sunrise, i.e. about from 4 to 7 a.m.

It wasn't much fun being on duty in this remote spot and in those conditions while the other fellows had all the Hawaii hotspots at their disposal and lived in air-conditioned barracks.

They had switched on and the radar was warming up. Lockard was keeping an eye on the hypnotising dance of the oscilloscope. He busied himself plotting the permanent interference, to hand in at the next weekly inspection – if the officer on duty didn't forget. Now and then Elliot asked if there weren't at long last any blips to see. Lockard did not even bother to reply. Blips? What blips? There were never any planes up, so there couldn't be any blips – especially on a Sunday!

The radiant sun dispersed the mist and rose over the peaceful, flower-decked island.

At 6.45 Lockard suddenly saw a very faint blip appear on the extreme edge of the screen, right at the top. Bearing 330 degrees. At 6.55 the blip, which seemed to be zigzagging slowly, coming back on itself and then going north again, became quite clear for a moment.

Lockard, without warning Elliot, who was busy outside with the generator, put a call through to Control over the normal line. After a five minutes' wait he was put through to a duty officer who took a dim view of the whole thing and told him pretty sharply to mind his own business.

Out of sense of duty Lockard made a note of the tracks and the times. At 6.58 the blip, still very faint, disappeared completely. At 7 o'clock, as per instructions, Lockard switched off and locked up the trailer. The two men stood waiting for the truck which was to take them back to the Mess for breakfast, when they heard the phone go. It was to tell them they couldn't get picked up till 7.30.

Lockard, furious, and having nothing better to do, went back to his radar and switched it on again, while Elliot put a shine on his shoes in a corner.

7.02 – 'Hi, George, come over here, quick!'

Elliot rushed over.

'Look!'

A miracle! For the first time for three months since they had been looking after that radar set, here was an honest-to-goodness blip, perfectly sharp. The big transparent green blip was moving fast southwards, towards Hawaii.

'Plot it!'

Elliot quickly placed a round sheet of transparent squared paper over the map, stuck a pin through the middle over where their station was, and got ready to take down the dope.

'7.02. Point 130. That was where they were when I first saw them!'

Probably Navy planes, thought Elliot, from an aircraft-carrier on manoeuvres out at sea, bringing ashore officers who wanted to spend a Sunday with their families. But they generally operated south of the island.

'Take down – 7.04 – 132.'

Elliot did a quick calculation on a special slide-rule. The planes were going about 225 m.p.h. Lockard kept his eye on the bright dot quivering on the screen and hesitated. Would he get bawled out again? He unhooked the special phone on the direct line to Control at Fort Shafter, reserved for urgent messages. Only after furiously winding the handle for several minutes did he hear the receiver being lifted at the other end. He recognized the voice.

'Hullo, Macdonald? Is that you, Joe?'

'Yeah.'

'Lockard here, from Opona Station. Find me someone at Central Control, it's very important.'

'There ain't no one here any more. The place shuts at seven, didn't you know?'

'Say, listen, Macdonald, be a pal and find me an officer, any one will do.'

Three minutes passed, while the blips got bigger and started to split up as they got nearer. Then a curt voice spoke in the phone:

'Lieutenant Tyler here, duty officer.'

'Lockard here, Lieutenant, with the SCR at Opona. I have spotted an important formation of planes making for Hawaii. At 0702 hours they were 136 miles away, and bearing zero to 10 degrees.'

'H'm' – The voice was perplexed and hesitant. 'O.K. I've got it. Don't worry. It's O.K.' The duty officer hung up.

At 7.28 the signs got lost in the fringe of permanent blips. The formation of aircraft was only twenty-two miles away. At 7.30 the truck picked up Elliot and Lockard, but they were fated not to get their breakfast that day.

At 7.57, on the glass-fronted balcony of the tall control tower on Hickham airfield, overlooking the naval base, Colonel Bertholf, who was worrying about the arrival of sixteen Flying Fortresses coming from San Francisco, suddenly leaped to his feet. In his binoculars he could see a long line of black dots – aircraft – approaching Kanai. There were about fifty of them, all single-engined.

The Colonel turned to the controller and asked him if they were Navy aircraft. Without even looking up he answered that it was unlikely, so early, on a Sunday morning.

A few moments later the planes began to peel off one by one and dive on Pearl Harbor. Showers of spray arose among the ships of the line jammed tightly together and powerless to move.

It was the Japs! *

At six in the morning Vice-Admiral C. Nagumo – commanding the First Air Fleet detailed for the attack on Pearl Harbor – had hoisted the 'Z' flag on the mast of the aircraft-carrier *Akagi*. It was the original flag hoisted by Togo at Tsushima in 1905 when he won

---

* In addition to the radar information, two other facts ought to have alerted Pearl Harbor. At 6.51 a Japanese midget submarine was sunk, just after coming into the fairway, by the destroyer *Ward*, which immediately sent in a signal. At 6.53 the control room on the aircraft-carrier *Enterprise* heard the pilot of one of its patrolling aircraft sent over Ford island, Ensign Manoel Gonzales, yell: 'Don't shoot. I'm American. Christ!' The Zeros were shooting him down.

Unfortunately no one was in command at Pearl Harbor. Admiral Stark and General Short each kept a jealous eye on the other Service, stuck to his own planes, and tried to palm off all the dirty jobs on to the other.

the first great naval battle of modern times, brought on board the flagship in a lacquer and gold casket.

'Z' – attack!

To fulfil his mission Nagumo had under his command six aircraft-carriers protected by nine destroyers, two heavy cruisers and two battleships. His six aircraft-carriers, with their romantic names, were: the *Akagi* (Red Castle), the *Shokaku* (Climbing Crane), the *Zuikaku* (Happy Crane), the *Kaga* (Increasing Joy), the *Soryu* (Green Dragon) and the *Hiryu* (Flying Dragon).

The squadron's senior navigation officer had surpassed himself, in spite of the bad weather which had prevented him from verifying his positions and his time schedule by the stars. At zero hour the exact position for launching the attack – lat. 26° N. long. 128° N. – had been reached.

The first wave comprised ninety Kates* (forty equipped with special torpedoes driven by oxygen, and the others with 1000-lb. bombs) and fifty Val dive-bombers, the whole escorted by fifty Zeros. This first formation was entrusted with the principal task, i.e. the sinking of the American battle fleet.

The second wave, whose task was to neutralize the airfields, comprised fifty Kates, eighty Vals and forty Zeros.

At 7.56 the war between the United States and Japan began.

In the officers' club at the great Wheeler airfield a poker game was just finishing. The previous evening – Saturday, 6th December – there had been a dinner-dance and the game had begun at 1 a.m., when most of the guests went home. At 6 in the morning play was still going on at one solitary table, and half an hour later two young Air Force Lieutenants, cleaned out to the last cent by three Navy officers, went out for a breath of air.

What was there to do so early on a Sunday morning? Lieutenants Welch and Taylor were muzzy with alcohol and cigarette smoke. They decided to have an invigorating bathe in the sea, followed by a few hours' sleep on the warm sands of Haleiwa beach by their airfield. Their car was the only one left in

---

* These code-names were not given until later in the war, but are used here for convenience. The planes were Aichi Type 99 (Val); Nakajima Type 97 (Kate); and Mitsubishi Type O (Zeke, generally known as Zero).

the car park. In front of them stretched the runways of Wheeler Field covered with aircraft. For a moment they admired the seventy-five magnificent Curtiss P-40s, just out of their crates, parked nose to nose and wing to wing to prevent possible sabotage by Japanese agents. In front of them two sentries were pacing up and down.

Just as Taylor was slamming the door of the Ford, bought second-hand and painted bright orange – they were young – a formation of planes swooped over the hangars with a thunder of exploding bombs. For the fraction of a second the young men sat paralyzed, but they sprang into action soon enough when a hail of bullets bespattered them with asphalt from the road. It was only then that they saw the red Japanese discs on the elliptic wings of one of the planes – a small squat monoplane with fixed undercarriage.

'Jesus, the Japs! – it's a dive-bomber!'

Welch backed his car viciously into the shelter of the club verandah and leapt to the telephone in the hall, while bullets were sending the tiles flying. He called up his unit.

'Get two P-40s ready – mine and Taylor's. Load up – it's not a gag, the Japs are here! Get going.'

He hung up, rushed out past the petrified Mess staff, who were cowering behind arm-chairs, tripped over a body and jumped into his car. Foot hard down on the accelerator, cursing his stupidity for having the car painted such a conspicuous colour, he roared along the twisting but luckily empty road at eighty miles an hour. The few cars he met were stationary, their occupants prudently lying in the ditches – including the traffic M.P.s apparently, as they saw three unattended red motorcycles propped up against some telegraph poles. No danger of a charge for speeding, anyway, and they took their corners on two wheels.

The nine miles took them less than ten minutes and on the way they were strafed – and missed – three times by the Japs. On the airfield they skidded to a halt on the damp grass and ran towards their planes. The fitters jumped down from their cockpits and a private staggered up with their parachutes, helmets and gloves. The engines were already running.

One minute later the two Curtiss's took off wing-tip to wing-tip and plunged into the low clouds coming in from the sea.

Haleiwa, tucked away at the northern extremity of the island in a hollow in the hills and covered by a providential layer of cloud, had miraculously escaped the attention of the Vals and Zeros, which had all been attracted by the enormous fires at Pearl Harbor and Hickham Field in the south. It is true that the well-informed Japanese knew it was a tiny training-field, where planes only came for shooting practice. As a matter of fact, 47 Squadron with its four P-36s and its fourteen P-40s had been there no more than a few days.

In all only seven fighters – and each one on his own initiative, like Welch and Taylor – managed to get into the air and intercept. Between them they brought down twelve Japs, whom they caught by surprise. Welch alone bagged four. However, each time they came up against Zeros instead of Vals they were outclassed. Lieutenants Christiansen, Whiteman, Bishop, Gordon Sterling, Dains and finally Taylor, Welch's friend, were all brought down in this way.

Lieutenant Welch, who managed to survive not only the Zeros but also the threat of a court-martial for having taken off without orders, received the D.S.C. a fortnight later. Subsequently, when the reports were gone through in Washington and General Arnold saw that Welch had taken off no fewer than three times – the last time, with two machine-guns out of four out of action, he had baled out – he recommended him for the Congressional Medal of Honour.

It was said at the time that the Air Officer Commanding in Pearl Harbor opposed the recommendation because Welch had taken off without orders. But finally, in spite of his outstanding war record, Welch was never awarded this high decoration.

At 10.15 on the 7th of December it was all over in Hawaii. Those responsible for the defence of the island got their stories ready, knowing that they would have to account for what had happened, and that scapegoats would have to be found for public opinion to tear to pieces.

Out of eight battleships in the harbour, five were sunk and three very badly damaged. At Kanehoe, the Naval air-base, out of thirty-six PBY Catalinas, large twin-engined long-range flying boats, twenty-seven were destroyed and six so seriously damaged as to be beyond repair. Only the three which were on patrol south of the island at the time of the attack escaped.

At Ewa, also a Navy air-base, there were eleven Wildcat fighters, thirty-two observation planes and six DC-3 transports. Fifteen Zeros, by machine-gun fire alone, in four minutes destroyed nine fighters, eighteen observation planes and all six transports.

The Army Air Force had, at 7 a.m., 221 good war-planes, dispersed on the three main airfields, Hickham, Wheeler and Bellows Fields, and the small satellite field at Haleiwa. Three hours later it had lost: eight B-17 Flying Fortresses, twenty-two B-18s and seven A-20s, all bombers. What was particularly serious was that every single fighter had been destroyed – sixty-two P-40Bs, eleven P-40Cs, twenty-three P-36s and nine P-26s. Every single plane, whether Navy or Army, which had escaped was damaged.

The Japanese had lost nine planes in the first wave and twenty-one in the second, a loss of thirty planes to put out of action 75 per cent. of the total American naval and air forces in the Pacific.

If the White House wanted the American people to be shaken out of its apathy, then it got what it wanted with a vengeance.

Clostermann's first book was his story of his own adventures in the war, serving with the RAF. From it comes this account of a battle late in the war between Spitfires and long-nose Focke-Wulfs of the Luftwaffe. It is simple, graphic, and vivid.

26

FROM: *The Big Show*

BY   Pierre Clostermann

Chatto & Windus, London, 1951

2nd July, 1944

'Scramble, south-east of Caen, as many aircraft as possible!'

Frank's shout tore us out of our lethargy. Great commotion! Where were the pilots? Were the aircraft ready?

Most of the pilots were having lunch and, as the squadron had just returned from a show, only a few of the aircraft were refuelled. I unhooked my helmet as I passed, looked for my gloves for a moment, then gave it up; as I hurriedly strapped on my Mae West I asked what wavelength was being used.

'Channel B! Hurry up, for Christ's sake!' shouted Ken who was already racing like a madman towards his plane. Luckily my old LO-D was ready, and my mechanics, who had heard the scramble siren, were already on the wing, holding out my parachute half done up. I put it on like a jacket while Woody started up my engine. I strapped myself in in a hurry. Three aircraft from Flight B were already taking off in a cloud of dust and Ken was waiting for me, with his engine ticking over, at the edge of the track. I took up my position and we took off.

Queer sort of weather, eight-tenths cloud at 3,000 feet, five-tenths at 7,000 and a great bank of stratus covering the whole of our sector as far as the Orne canal. At 12,000 feet, there was a ten-tenths layer of nimbo-stratus. Ken and I managed to catch up with Frank, the Captain and Jonssen the Norwegian over Caen at 6,000 feet. Control gave us vague courses to patrol, and told us to keep our eyes open for two unidentified aircraft moving about in the clouds near us. We climbed to 7,000 feet, just on a level with the second cloud-layer. In the distance, out of range, a few suspicious black dots were moving among the cumulus. Suddenly Frank's voice sounded in the earphones:

'Look out chaps, prepare to break port.' I went into a slight left-handed turn and looked up. A solid mass of forty German fighters were emerging from the clouds 3,000 feet above us. We couldn't identify them yet – Messerschmitts or Focke-Wulfs – but one thing was certain, they were Jerries. The way they flew was unmistakable. The nervous waggling of the wings, their, at first sight, untidy formation. A heady feeling of elation swept over me and my hand trembled so much that I only succeeded in taking off the safety-catch at the third attempt. I felt on top of my form. Instinct schooled by long training functioned smoothly; I tightened my safety-straps, huddled down on my seat and shifted my feet up the rudder bar. Excitement keyed my muscles to their highest pitch of efficiency, all fear vanished. My fingers were in harmony with the controls, the wings of my plane were extensions of myself, the engine vibrated in my bones.

I began to climb in a spiral. Now! The first Huns released their auxiliary tanks, fanned out and dived towards us.

'Break port! Climbing!' Full throttle, 3,000 revs. a minute, we faced the avalanche.

They were Focke-Wulf Ta 152's. My Spitfire was climbing at 45°, hanging on the propeller. I intercepted the first group, which was diving in line ahead on Frank's section. Frank made the mistake of diving for the clouds, presumably to gain speed but forgetting the vital principle: 'Never turn your back on the enemy'. As we crossed I managed to get in a burst on the front Hun whose wing lit up with the explosions. Three or four puffs of white smoke appeared in his slipstream. Two 'long-nose' Focke-Wulfs did a tight turn bringing them head-on to me, and the tracer from their 20-mm. Mauser 151's formed long glittering tentacles snaking towards me and curling down just under my fuselage. The sky began to be a whirling kaleidoscope of black crosses.

In a dog-fight at over 450 m.p.h. you sense rather than see the presence of aircraft circling round, until suddenly your eyes fix on one of them.

I fixed one of them now! A Focke-Wulf. He was circling, his

black crosses edged with yellow and his cockpit glittering in the sun. He was waggling his wings looking for an opponent too. Now I had him framed in my sights. Ought I fire? Not yet. Patience . . . still out of range. But he had seen me, fell off to starboard and went into a tight turn. Two white 'contrails' appeared at his wing-tips. He then began a vertical climb, straight up like a rocket. Suddenly he turned on his back with such violence that in spite of his change of attitude, his momentum continued to project his glittering belly towards the sun. Within range at last! I jammed my thumb down on the button and my wings shuddered with the recoil of the cannon. With one motion of the stick I made the luminous spot of my gunsight travel along the Hun, through his propeller slowly churning through the air like some pathetic windmill. I was now so close to him that every detail was clear. It was one of the latest 'long noses' with a Daimler-Benz in-line engine. I could already see the little blue flames of the exhausts, the oxide tail left by the burning gases along his fuselage, his emerald green back and his pale belly like the pike I used to fish for in the old days in the Mayenne. Suddenly the sharp clear picture shook, disintegrated. The gleaming cockpit burst into fragments. My 20-mm. shells tore into him, advancing towards the engine in a series of explosions and sparks that danced on the aluminium. Then a spurt of flame, thick black smoke mingled with flaming particles. I must get out of the way. I put all my weight on the controls and as my Spitifre flicked off I had a last vision of the Focke-Wulf, disappearing down below like a comet towards the shroud of clouds covering the Orne canal.

The whole thing had hardly lasted a few seconds. Never before had I felt to the same extent the sudden panic that grips your throat after you have destroyed an enemy aircraft. All your pent-up energy is suddenly relaxed and the only feeling left is one of lassitude. Your confidence in yourself vanishes. The whole exhausting process of building up your energy again, of sharpening your concentration, of bracing your battered muscles, has to be started all over again. You would be glad to escape, you hurl your aircraft into the wildest manoeuvres, as if all the German fighters in the entire Luftwaffe were banding together, and concentrating their threat exclusively on you. Then the spark strikes again, the partnership of flesh and metal reforms.

163

To my right a Spitfire broke off and dived behind a Focke-Wulf. I caught a glimpse of the markings – LO-B; it was Ken. I must cover him and, avoiding several determined attacks by Huns, I went into a tight spiral dive – they were moving too fast to follow me.

Ken fired; his wings disgorged long trails of brown smoke and a rain of empties. Intent on watching him I was paying no attention to anything else. A shadow formed, covering my cockpit. I looked up. Thirty feet above, a Focke-Wulf's enormous oil-bespattered belly passed me. He had missed me and opened fire on Ken.

Instinctively I throttled right down, pulled gently on the stick, aligned him in my sights and, at point blank range, opened fire. The stream of steel belched forth by four machine guns and two cannon smacked into him at 150 yards range just where the starboard wing joined the fuselage. The Focke-Wulf, shaken in its course, skidded violently to the left and the right wing folded up in a shower of sparks, parted company with the fuselage, smashed the tailplane and whizzed past me in a hail of fragments.

I had scarcely recovered from my surprise when six other Focke-Wulfs attacked me. Turmoil ensued and I defended myself like one possessed. Sweat poured off me and my bare hand slipped on the stick.

Three thousand feet above, Frank's section was trying to hold its own in a whirling mass of Focke-Wulfs.

The only thing to do was to keep constantly turning, while the 152's stuck to their usual tactics – diving attacks followed by vertical climbs. We had one factor on our side; we were fighting fifteen miles from our base, while the Huns were 150 from theirs. They would be the first to pack up.

All the same, I got a bit fed up with this rigmarole. I succeeded in nabbing one who was slow in straightening out from his climb: shells exploded under his belly. The usual spin, the usual tail of thick black smoke. It would have been hazardous to go down after him – I should immmediately have had half a dozen others on top of me. Oh well, he wouldn't be a 'certain' but I should be satisfied to have him put down as a 'probable'.

No time for repining, anyway. Other things to think about. My port cannon jammed. I pumped my remaining twenty or so starboard shells into a Focke-Wulf whom I caught doing an impeccable roll. What an extraordinary idea to do a roll in the middle of a scrap! As the British say, there is a time and place for everything!

The Focke-Wulfs seemed to have had enough and showed signs of weakening. Apart from three or four who continued to attack, they set course south. I took the opportunity of sidling discreetly off to the clouds. I was exultant, for, in forty minutes, I had scored three successes, two of which would be confirmed, and I had damaged two other planes. I indulged in the luxury of five victory rolls over Longues, to the joy of the country folk.

Technology took a large part in the air war, with our boffins working at fiendish devices to counteract the devices of their scientists, and vice versa. The Germans were not slow after the Battle of Britain to appreciate the power of radar, and soon had a radar and fighter control defence system to match ours. The ground radar stations would plot the Allied bombers as they came on and radio details to the German night fighters who would then try and close on them close enough to pick up a bomber with their own primitive and short-range airborne radar sets.

Here one of the leading Luftwaffe night fighter pilots describes a raid in which the British gained a brief technological history by using Window – radar reflecting silver paper chaff. Then he quotes his commanding officer, an older warrior, moralising on the futility of war, and suggesting that all that combatants achieve is to destroy themselves. Not only were They every bit as brave as Us, they could be as thoughtful, too.

27

FROM: *Duel Under the Stars*
BY   Wilhelm Johnen
William Kimber, London, 1957

On the 27th July, 1943, there was something in the air. The early warnings from the Freya apparatus on the Channel coast indicated a large-scale British raid. In the late afternoon various flak units, night-fighter wings and civilian air-raid posts had been given orders to have their full complement at action stations. What were the British up to? What city that night would be the victim of these well-prepared raids? Every ominous presentiment was to be fulfilled that night. In all ignorance, the night-fighter squadrons took off against the British bombers, whose leaders were reported over Northern Holland.

I was on ops and flew in the direction of Amsterdam. On board everything was in good order and the crew was in a cheerful mood. Radio operator Facius made a final check and reported that he was all set. The ground stations kept calling the night fighters, giving them the positions of the bombers. That night, however, I felt that the reports were being given hastily and nervously. It was obvious no one knew exactly where the enemy was or what his objective would be. An early recognition of the direction was essential so that the night fighters could be introduced as early as possible into the bomber stream. But the radio reports kept contradicting themselves. Now the enemy was over Amsterdam and then suddenly west of Brussels, and a moment later they were reported far out to sea in Map Square 25. What was to be done? The uncertainty of the ground stations was communicated to the crews. Since this game of hide-and-seek went on for some time I thought: To hell with them all, and flew straight to Amsterdam. By the time I arrived over the capital the air position was still in a complete muddle. No one knew where the British were, but all the pilots were reporting pictures on their screens. I was no exception. At 15,000 feet my sparker announced the first enemy machine in his Li. I was delighted. I swung round on the bearing in the direction of the Ruhr, for in this way I was bound to approach the stream. Facius proceeded to report three or four pictures on his screens. I hoped that I should have enough ammunition to deal with them!

Then Facius suddenly shouted: 'Tommy flying towards us at a great speed. Distance decreasing . . . 2,000 yards, 1,500 . . . 1,000 . . . 500 . . .'

I was speechless. Facius already had a new target. 'Perhaps it was a German night fighter on a westerly course,' I said to myself and made for the next bomber.

It was not long before Facius shouted again: 'Bomber coming for us at a hell of a speed. 2,000 . . . 1,000 . . . 500 . . . He's gone.'

'You're crackers, Facius,' I said jestingly.

But I soon lost my sense of humour for this crazy performance was repeated a score of times and finally I gave Facius such a rocket that he was deeply offended.

This tense atmosphere on board was suddenly interrupted by a ground station calling: 'Hamburg, Hamburg. A thousand enemy bombers over Hamburg. Calling all night fighters, calling all night fighters. Full speed for Hamburg.'

I was speechless with rage. For half an hour I had been weaving about in a presumed bomber stream and the bombs were already falling on Germany's great port. It was a long way to Hamburg. The Zuider Zee, the Ems and the Weser disappeared below and

167

Hamburg appeared in the distance. The city was blazing like a furnace. It was a horrifying sight. On my arrival over the city the ground station was already reporting the homeward flight of the enemy in the direction of Heligoland. Too late! The flak gunners had already ceased to fire and the gruesome work of destruction had been accomplished. In low spirits we flew back to our airfield.

How could the German defences have been rendered so impotent? We know today. The British had procured an example of our successful Li apparatus and had found the countermeasure. With ridiculous strips of tinfoil they could now lure the entire German night-fighter arm on to false trails and reach their own target unmolested. It was a simple yet brilliant idea. As is well known, radar works on a determined ultra-short wave frequency. By dropping these strips of tinfoil the British jammed this frequency. In this way the air goal was achieved and for the night fighter the bomber had once more become as invisible as it had been before the invention of the Lichtenstein apparatus.

While the main bomber stream far out to sea was flying towards Hamburg, smaller formations had flown over Holland and Belgium to Western Germany, dropping millions of tinfoil strips. These 'Laminetta' appeared on the German screens as enemy bombers and put various ground stations out of action. The smaller formations, according to schedule, next dropped enormous quantities of flares – the famous Christmas trees – over various cities in the Ruhr. A few bombs were also dropped. The night fighters streaked towards these signs of attack from all directions, looking in vain for the bomber stream.

In the meantime the leaders of the British main raiding force reached Heliogoland unhindered and dropped more strips, putting the ground detectors out of action. At one blow both ground and air defence had been paralysed. In daylight on the following morning, whole areas of Holland, Belgium and Northern Germany were strewn with these strips of foil. Certain people maintained that they were poisonous and that they would kill the cattle. The innocuousness of these small pieces of tinfoil on the ground was soon apparent, but in the air they were deadly – fatal for the life of a whole city.

A few days later we heard further details of the agony of this badly hit city. The raging fires in a high wind caused terrific damage and the grievous loss of human life outstripped any previous raids. All attempts to extinguish them proved fruitless and technically impossible. The fires spread unhindered, causing fiery storms which reached heats of 1,000°, and speeds approaching gale force. The narrow streets of Hamburg with their countless backyards were favourable to the flames and there was no escape. As the result of a dense carpet bombing, large areas of the city had been transformed into a single sea of flame within half an hour. Thousands of small fires joined up to become a giant conflagration. The fiery wind tore the roofs from the houses, uprooted large trees and flung them into the air like blazing torches. The inhabitants took refuge in the air-raid shelters, in which later they were burned to death or suffocated. In the early morning, thousands of blackened corpses could be seen in the burned-out streets. In Hamburg now one thought was uppermost in every mind – to leave the city and to abandon the battlefield. During the following nights, until the 3rd August, 1943, the British returned and dropped on the almost defenceless city about 3,000 blockbusters, 1,200 landmines, 25,000 H.E., 3,000,000 incendiaries, 80,000 phosphorus bombs and 500 phosphorus drums; 40,000 men were killed, a further 40,000 wounded and 900,000 were homeless or missing. This devastating raid on Hamburg had the effect of a red light on all the big German cities and on the whole German people. Everyone felt it was now high time to capitulate before any further damage was done. But the High Command insisted that the 'total war' should proceed. Hamburg was merely the first link in a long chain of pitiless air attacks made by the Allies on the German civilian population.

Shortly after our return from the West to Parchim, we had a little celebration, where the drink flowed freely. A fellow pilot from Essen, Peter Spoden, who was very popular in the mess, had to keep going down to the cellar. The 'old man', Hauptmann Schönert, was on free and easy terms with the whole mess. He remembered his own youth and spoke enthusiastically of his life as a sailor. During a lull in the party he suddenly stood up, grabbed one of us by the shoulders and dragged him to the window. The last rays of the evening sun were breaking over the cloud banks on the horizon and staining the heavens scarlet. The sky was a riot of colour, from the most delicate blue to fiery red on the rising cloud bank. The evening had fallen and deep peace lay over the land. The C.O. opened the window to let in the cool air. The pine woods were fragrant. He puffed his cigar with relish and turned to us with a smile.

'Boys, you've bitten off a pretty hard chunk. You get into your crates and are swallowed up in the darkness. Some of you return from single combats at thousands of feet above the tortured, burning earth. Each of you flings his life without a murmur into the scales. A bloody hard life.

'At the age of twenty I lived a carefree happy one. I sailed the seven seas as a ship's boy and learned to appreciate and love other nations. You could find good pals everywhere. We were all flung together in our cargo boats – Britishers, Norwegians, Danes and Germans. At first the atmosphere was very cold, but once we had the first storm behind us, we smiled at each other. During the first days at sea it was each man for himself, and yet as soon as we began to feel homesick we grew closer to each other. Soon we were pals and brothers. To hell with all prejudices! Here in the howling storm, when the huge breakers washed the decks and Father Death stood in the bows, was to be found the real League of Nations. We laughed at the raging seas.

'And this laugh meant confidence and mutual aid in life and death. When the gale blew itself out we had a breathing space and we had been granted a new lease of life. On reaching our home port after many months, we had become a community which recognised no difference of people, race or speech. With heavy hearts we said farewell in the hope that we should never forget comrades who had shared our joys and griefs.

'I have found the same comradeship among you. We, too, are faced with the same dangers and yet . . . there's something eating me.' There was a bitter look on his face as he said these words. 'We are destroying ourselves. Our fight is not against the powers of nature for the good of humanity but an attempt to destroy life with all the new weapons of science.

'Do not men of our race – perhaps the fair-haired Britishers with whom I sailed in the Bay of Biscay and made friends – sit in their bombers, night after night, turning our cities to ruins? Each does his duty. But don't we thereby aggravate our hatred? At night we see only the enemy bomber and its bright red, white and blue circles. Our cities burn. The bomber must be brought down at all costs, and when it crashes we crow. We only see the bomber burning and not the crew. We only see the emblem laid low, not the youngsters hanging on their straps in their death agony.

'And then perhaps one day you meet a Tommy who has baled out. You meet him down below. His eyes have lost the harsh glint of battle. You shake hands and this handshake is the beginning of a comradeship, born of a life and death struggle. Gratefully he accepts the cigarette you offer. The barrier that divided us has fallen and two men stand facing each other. Hostility and propaganda have made them enemies but the common danger of battle has made them friends. Just as here, in a small way, hatred is changed to friendship, may the racial hatred also turn to friendship. But the iron carapace in which the nations shroud themselves, the outward symbols of which are emblems and threats, must be swept away, for the more the modern world uses science, the bloodier will the battles become. The more man takes refuge behind armour plating and steel, the greater will be his will to destruction. For this reason this bloody murder must come to an end. The people must lay aside their blood-stained armour unless the whole world is to be destroyed. All the peoples of the world could live in peace, and this path must be taken together and protected so that they could all rise in judgement against anyone who left the path. . . .' Hauptmann Schönert looked up at the darkening sky.

Short indeed were these hours of relaxation for the menace of the British hung like a sword of Damocles over the German cities. They were trying to destroy the heart of Germany from the air. All of us were living under the spell of the approaching catastrophe. This rest period had to be used for training the newcomers. Night after night we flew, practised and trained, until the recruits could handle their machines with the precision of a sleepwalker.

General Curtis E. LeMay is the man who built America's Strategic Air Command bomber and missile force – the mightiest armada there was. An unlikely army: 'Peace is our profession' is their motto, and if they did ever have to be launched, they would have failed in their primary role – as a deterrent. The legendary LeMay was a four-star general for longer than any other in American history. His was the responsibility for the dropping of those atomic bombs in Japan in 1945; that this is still the only offensive use of nuclear weapons ever made must be due largely to the existence of SAC.

LeMay conceived the awesome low-level fire bomb raid by B-29s that burned in Tokyo in March 1945. He had earlier led the historic and horrific B-17 1943 attack on the German city of Regensburg in which the USAAF lost 60 out of 307 bombers – that's one in five. He had evolved the huge formations that enabled the B-17s, the Flying Fortresses, to concentrate their defensive fire. Even earlier, as a boyish lieutenant, he had been lead navigator in a mass formation of bombers flying down non-stop to Buenos Aires in 1938 on a goodwill and training mission.

General LeMay's autobiography reveals him not as a crusty old warrior of legend, but as a sensitive and imaginative man whose courage was all the more admirable for the force of the anxieties which beset him when he came to make those awesome command decisions. The book offers a compelling, vivid insight into the mind of one of the great commanders of history.

I have chosen two excerpts: first, LeMay's own account of the disastrous Regensburg raid; and second, his account of his thinking while deciding to send his B-29 Superfortresses over Tokyo by day at low level.

28

FROM: *Mission with LeMay*

BY General Curtis E. LeMay with MacKinley Kantor
Doubleday, New York, 1965
Regensburg

On this fateful Tuesday, August 17th, we in the Third Division got together all right, delayed a bit by weather. Possibly we were ten or fifteen minutes behind the original schedule. We wondered what was happening to our friends over in the First.

Had we known the truth, we would have thought that we should have stood in bed. The First Division just wasn't getting off the ground.

Let me try to clarify the plan of battle once more. I was going in first with my division, and we would fly right straight through and hit Regensburg, and go on out through the Alps, down to the Med and to Africa. Ten minutes behind us, the (elder and vastly more numerous) First Air Division, commanded by General Bob Williams, was to proceed on their attack against Schweinfurt. (That wasn't as deep a penetration as ours, not by over a hundred miles.) They'd hit Schweinfurt, then turn around and fly back to England.

In our division we'd have the benefit of fighter support as far as the range of the fighters would permit them to go, inland over the Continent. Then, as Williams came out, they would pick him up, and cover him on the way out.

I flew in the first aircraft of the 96th Bomb Group. In all we dispatched a hundred and forty-six airplanes to strike Regensburg, and a hundred and twenty-seven of these succeeded in attacking the primary target. But I'm getting ahead of myself.

Here we were, up in the air, all assembled, and there were no fighters around, and there wasn't any First Division airborne. Old Fred Anderson down there at Pinetree was in severe trouble. He had to decide whether to scrub the whole mission, or send me in alone. So he said Go. We went.

Actually what he was doing was holding up the Schweinfurt people for three and a half hours. Their weather was stinko, as ours had been.

Folks tried to look on the bright side: (a) the weather was due to improve; couldn't get any worse; and (b) they would have the benefit of fresh fighter cover. The fighter boys could finish escorting us, and have time to return to England, land, refuel, and get airborne once more to offer their best possible protection to the First Division. (Still weren't enough belly-tanks in Fighter Command; there were only a few around. And their range couldn't be extended without belly-tanks.)

Now here's what seems apparent to me about that mission:

If the First Division had been concentrating on the same sort of bad-weather-instrument-take-off procedure which we had been

173

developing for a solid month, they might have been able to get off the ground as we in the Third did. A few minutes late, perhaps; but still part of the originally-planned show.

And we couldn't horse around about this – return to our bases, sit on the ground, take off once more – even if weather permitted. We had to land in Africa before dark. It's a long way way to Africa from England. I was faced with the unhappy notion that we might drop at least three or four 17's down there in the Mediterranean – just because they were out of gas.

Yep. We did.

The history books will tell you that eighteen squadrons of Thunderbolts (P-47's) and sixteen squadrons of Spitfires were assigned to provide cover for our bombers on that day. And I will tell you that I led that mission, and not one damn Jug (P-47) or one damn Spit did I see.

Our fighter escort had black crosses on their wings.

I lost twenty-four out of my hundred and twenty-seven planes which attacked the target at Regensburg. Hours later the First Division lost thirty-six bombers, but they had one hundred and eighty-three planes in the battle. That made a total of sixty which went down that day. The previous high had been suffered when we went to Bremen, 13 June: the Eighth Air Force lost twenty-six.

Four or five of my crews managed to limp over to Switzerland and they got down more or less safely there. The rest, in the B-17's which had become casualties, were either Prisoners of War or they were dead.

A friend wrote me a descriptive letter, a while after that mission.

He said, 'That's what I still can't get out of my mind. There were two different 17's which went *whuff*. That was it: just *whuff*, and they were gone. We saw debris flying around from one and saw absolutely nothing from the other. The plane and its entire crew and bomb load and everything else, seemed to disappear as if some old-fashioned magician had waved a wand.

'There were Forts falling out of formation with bad fires. Then we'd try to count the chutes, and then our attention would be directed somewhere else and we couldn't count any more chutes. We'd look again, and the airplanes were gone, and also the people

174

with chutes. There were more fighters coming in, with the leading edges of their wings all fiery. . . .

'And there was that one airman going down, doubled up, just turning over and over. He went right through the formation, and nobody seemed to hit him, and he didn't seem to collide with any of the airplanes. He just fell fast and furious, over and over, no chute, no nothing. I wondered who he was. Did he come out of a Fort named *Lewd Lucy* or one named *Wayfaring Stranger* or the *Nebraska Cornball*? I had friends flying in crews of Forts like that, and maybe he was one of my friends. But his own mother wouldn't have known him then, and certainly she wouldn't have known him after he hit the ground.

'As we got on the target it appeared that the flak was as nothing to the flak over the Ruhr or over a lot of other targets – Kassel, for instance. Keroway took off his mask to shake some water out of the hose, and he leaned across to me before he put it on again, and he said, "I was down there once on a tour with a bunch of students. They've got awful good beer. Will you have light or dark?" I told him I'd have both.

'Regensburg was the Ratisbon of that Robert Browning thing we had to learn in school. Think back now, to a certain very good play, *Life With Father*, in which a red-headed boy recited this poem on stage. And when he said *the boy fell dead* by gosh the little boy who was reciting fell down as if dead.

'Funny that the same town could be both Regensburg and Ratisbon; but it all depends upon where you went to school, I guess – whether you were born under a tricolor or under German eagles.

> You looked twice ere you saw his breast
>   Was all but shot in two. . . .
> 'You're wounded!' 'Nay,' his soldier's pride
>   Touched to the quick, he said:
> 'I'm killed, Sire!' And his chief beside,
>   Smiling, the boy fell dead.

'You know, that night after we landed down there in Africa, I kept hearing those words over and over. *Smiling, the boy fell dead.* Then I'd see that hunched-up figure, upsy-daisy, over and over, coming down through the formation. I guess he fell dead all right, there on a little mound near Ratisbon.'

175

Tokyo

. . . Must have had the idea in the back of my brain when I talked to Norstad. I kept toying with it. I said to Norstad, 'You know General Arnold. I don't know him. Does he ever go for a gamble? What do you think?'

. . . Let's see: we could load with E-46 clusters. Drop them to explode at about two thousand feet, say, or twenty-five hundred. Then each of those would release thirty-eight of the M-69 incendiary bombs. . . . Wouldn't have to employ all the same type of incendiaries, of course. Could use both napalm and phosphorus. Those napalm M-47's.

. . . They say that ninety per cent of the structures in Tokyo are built of wood. That's what Intelligence tells us, and what the guidebooks and the *National Geographic* and things like that – They all say the same. Very flimsy construction.

. . . *General Bush, sir. I'd be perfectly satisfied just to enlist, in order to get up on the priority list. Because I'm determined to go to flying school.*

. . . Bringing those 29's all the way down from thirty thousand to about nine or even five thousand. A lot of people will tell me that flesh and blood can't stand it. Maybe they'll be right: maybe flesh and blood *can't.*

. . . Norstad didn't have an idea what I was thinking about. But he did opine that he thought General Arnold was all for going in and getting the war won. Certainly Larry didn't say enough to convince me that I'd get off scot free if I made a mistake. But I did gain the impression that being a little unorthodox was all right with Hap Arnold.

. . . So this is what you call being a little unorthodox? What are you – British? Elveden Hall gone to your head? Want to be president of the Department of Understatement? O.K. You're elected.

. . . I think it was Velma – maybe it was Lloyd – who had that toy village. Well, maybe some other kid in the neighbourhood really owned it. But it was one of those villages that you set up . . . the houses come all flat, but they're hinged at the corners; and then you spread 'em out and shove the roof down, with the eaves going up through slots and so on; and thus your house sits like a strawberry box. Well, I remember they had the village all set up out in the backyard. And some mean kid says, 'Let's see if we can burn it down.' So he set fire to the first house. And, brother, they all went.

. . . Just which Medium outfit was it? I know it was over in England, and it was Mediums. Yep. The 322nd, along in May of '43. I think they had one abortion. But the rest of the group went on to Holland and attacked at low altitude. Not one B-26 came home that day. So that's what happens at low altitudes.

. . . Always?

. . . The mother writes you a letter, and she says: 'Dear General. This is the anniversary of my son Nicky being killed over Tokyo. You killed him, General. I just want to remind you of it. I'm going to send you a letter each year on the same date, the anniversary of his death, to remind you.'

. . . All right, by God. If I do it I won't say a thing to General Arnold in advance. Why should I? He's on the hook, in order to get some results out of the B-29's. But if I set up *this* deal, and Arnold O.K.'s it beforehand, then he would have to assume some of the responsibility. And if I don't tell him, and it's all a failure, and I don't produce any results, then he can fire me. And he can put another commander in here, and still have a chance to make something out of the 29's. This is sound, this is practical, this the way I'll do it: not say *one word* to General Arnold.

. . . No bomb-bay tanks either. Nothing but bombs in those bomb-bays. We won't need all that extra gas if we're not going to altitudes.

. . . How do I get that way, saying *we*? *I* can't go on this mission if we run it. A man came and talked to me, and I know something about a Firecracker. So I can't go. No use asking, no use trying to pry loose the permission. It won't work. I know about the Firecracker now, and no other one of my people does know about it. . . . Tommy Power is all in favour of this low-level incendiary attack. If we run it, I'll let him lead.

. . . Just because a few people are for it, that doesn't mean that *everybody* is for it. And if we go to hell in a handbasket, *nobody* will be for it. When you dream up something successful, everybody else wants a part of it. They knew all along that it wouldn't work.

. . . At least we could say forget the weather. We've proved that even the stupidest radar operators can get us over that land-water contrast up there at Tokyo. If we send some veterans in ahead, they're bound to get on the target, and they're bound to start the fires. If we really get a conflagration going, the ones that come in later can see the glow. They can drop on that.

. . . *LeMay, what's the weather report for today? LeMay, how many airplanes have we got in commission today?*

. . . And there's another way we can save weight. Take out all the defensive armament. Repeat: remove all defensive armament from every B-29. Every gun goes out, all the ammunition goes out. No guns, no ammunition. At least our folks won't be shooting at each other.

. . . So if we go in low – at night, singly, not in formation – I think we'll surprise the Japs. At least for a short period of time; but certainly they will adjust to our new tactics before long. But if this first attack is successful, we'll run another, right quick. Say twenty-four hours afterward. Two days at the most. And then maybe another.

. . . 'Dear General Burnside. This is the anniversary of the death of my son Sam, whom you killed at Fredericksburg. Dear General Hancock. Twenty years ago today, you killed my son Benjy at Chancellorsville. I just want you to remember this, I will send you a letter every year.'

. . . Not only take out the guns and the ammunition. Take out all the gunners too. Less weight, and fewer people jeopardized. But I think we'd better leave the tail gunners in, for observation purposes.

. . . Let's see. Say we use an intervalometer setting of fifty feet, I reckon one aircraft could burn up about sixteen acres of territory.

. . . Plenty of strategic targets right in that primary area I'm considering. All the people living around that Hattori factory where they make shell fuses. That's the way they disperse their industry: little kids helping out, working all day, little bits of kids. I wonder if they still wear kimonas, like the girls used to do in Columbus in those Epworth League entertainments, when they pretended to be Geisha girls, with knitting needles and their grandmother's old combs stuck in their hair.

. . . Well, the new maintenance setup worked. And the Lead Crew school worked. And the critiques after the missions worked. And the combat box worked; and the straight-and-level bomb run worked. But that doesn't mean that everything I decide to do is going to work. You can't disregard the law of averages. Or can you? One of these fine days I'm going to dream up something that doesn't work, and this just might be the time.

    . . . LeMay, I want the Utah. *You'd better find it for me.*

They taught me how to fly,
And they sent me here to die.
I've got a bellyful of war. . . .

Come on and get promoted
As high as you desire.
You're riding on the gravy train,
When you're an Army flyer.
But just about the time you are
A general, you will find
Your wings fall off, and your engine quits,
And you will never mind.

    . . . We don't *think* that their night fighters amount to anything. We don't think that their *night fighters* amount to anything. We don't think that their night fighters amount to *anything*. And we could be wrong as hell.

    . . . If we try this, I want to try it with at least three hundred airplanes. Let's see: take the 73rd. I guess Rosie O'Donnell can put up at least a hundred and fifty, over there on Saipan. The 313th . . . I guess Skippy Davies ought to be able to put up at least a hundred; maybe more. The 314th, here on Guam, they're green as gourds. And they're not up to strength. I guess maybe Tommy could put up fifty.

    . . . Need at least three hundred over that target to do the kind of job that should be done. I wish to God we could send five hundred B-29's instead.

    . . . We can have some graphs drawn: indicate the experience which each crew has. That would help a lot in determining their gas loadings. Green crews always burn a lot more gas. An experienced crew can get the full range with a full bomb load. But an inexperienced crew always uses more gas.

    . . . Yes, and I had one grand chore convincing the crews that they ought to come down (from twenty-eight or thirty thousand feet) to twenty or twenty-three, for daylight stuff. I finally got that done, but we still haven't achieved any special results. Now I'm speculating on telling them that they've got to come down to five or seven thousand feet. And no gunners, and no guns.

. . . *Why all the eagerness to enlist in one of my batteries, son?*

. . . With at least three hundred planes we can get a good concentration. So that'll be our first mission – with all those sorry radar operators as well as the capable ones. Just go up and fly over that wad of land in Tokyo Bay, turn on the heading we give them, continue so many seconds, and pull the string.

. . . 'Dear General Washington. Dear General Knox. Dear General Gates. Dear General Greene. This is the anniversary of the day you killed my son Eben, my son Jeremiah, my son Watson, my son John. You killed him at Princeton, at Monmouth, at Saratoga, at German-town.'

. . . How many times have we just died on the vine, right here on these islands? We assembled the airplanes, assembled the bombs, the gasoline, the supplies, the people. We got the crew set – everything ready, to go out and run the mission. Then what would we do? Sit on our butts and wait for the weather.

. . . So what am I trying to do now? Trying to get us to be *independent of weather*. And when we get ready we'll *go*.

. . . Intelligence says that every one of those factories is surrounded by a hundred-foot firebreak. But if we really got rolling with incendiaries, and had a wind to help us with the flames, firebreaks wouldn't make any difference.

. . . Ninety per cent of the structures made of wood. By golly, I believe that Intelligence report said ninety-*five!* And what do they call that other kind of cardboard stuff they use? *Shoji.* That's it.

. . . I remember running those missions for Chennault out of the Chengtu. He was always pestering me to do things for him. I ran that mission against Nanking, and the one against Hankow. Incendiary attacks: warehouse and supply installations at Nanking, and docks and storage facilities at Hankow. Everything was fouled up there, beginning with Chennault's last minute request for a change in our Time over Target . . . people dropped in the wrong sequence, smoke obscured the primary areas, so on. But that was an *incendiary* attack, and everything which was hit burned like crazy. And I think there was a vast similarity to the *type of construction* in Japan.

. . . That's over three thousand miles. And no gas in the bombbays. Let's see: I think we could average out at about six tons of incendiaries per airplane.

. . . *You concern yourself with staying in formation, and let someone else worry about the flak.* Well, I'm worrying about it now.

'Dear General Marshal Ney. This is the anniversary of the day you managed to kill off my son François. . . . Dear General Pompey. This is the anniversary of the day you slew my son Junius. Dear General. . . .'

. . . R-Roger. How many types do we employ? We use the E-46 clusters, but we don't load entirely with that stuff. We've got to use some other types as well. Also mix in some high explosive bombs, especially in the 314th if I have Tommy lead. Those HE's will make the Japs stay under cover, and not come rushing out to extinguish the first fires.

. . . Each type of weapon has some good points as well as some bad points; but if I now had my choice, and had available an overwhelming quantity of any type of fire bomb which could be employed, I wouldn't stick to one particular type. No. Of course magnesium makes the hottest fire, and it'll get things going where probably the napalm might not. But the napalm will splatter farther, cover a great area. We've got to mix it up. We're not only going to run against those inflammable wooden structures. We're going to run against masonry too. That's where the magnesium comes handy.

. . . No matter how you slice it, you're going to kill an awful lot of civilians. Thousands and thousands. But, if you don't destroy the Japanese industry, we're going to have to invade Japan. And how many Americans will be killed in an invasion of Japan? Five hundred thousand seems to be the lowest estimate. Some say a million.

. . . We're at war with Japan. We were attacked by Japan. Do you want to kill Japanese, or would you rather have Americans killed?

. . . *I hope you're right, Curt.*

. . . Crank her up. Let's go.

Drafts from the Tokyo fires bounced our airplanes into the sky like ping-pong balls. A B-29 coming in after the flames were really on the tear would get caught in one of these searing updrafts. The bombers were staggered all the way from five to nine thousand feet, to begin with. But when fires sent them soaring, they got knocked up to twelve and fifteen thousand feet.

According to the Tokyo fire chief, the situation was out of control within thirty minutes. It was like an explosive forest fire in dry pine woods. The racing flames engulfed ninety-five fire engines and killed one hundred and twenty-five firemen.

Well, I told about it in the first chapter of this book . . . burning up nearly sixteen square miles of the world's largest city. I walked the floor down there on Guam all night long. Tommy Power (following my instructions) after he'd dropped his own bombs, flew back and forth over the scene, making pictures. He went up to twelve and even twenty thousand feet to examine the situation. He'd brought along some cartographic types who could sketch accurately; and unless I'm severely mistaken he made of lot of sketches himself. He reported to our HQ by radio, and told of the inferno. But – point is – I wouldn't know what our losses were until all the surviving planes were back on Tinian, Saipan, Guam.

. . . There were some photographs too, snapped in the glare of the flames, but of course they didn't compare with the candid record of next day, when reconnaissance planes flew over Tokyo. Fact is, fires were so intense that they didn't linger in the burning. The blaze was practically out by noon of March 10th. Then some very clear recco photographs could be taken. These I sent along promptly to General Arnold.

If it hadn't been for that big river curving through the metropolitan area, a lot more of the city would have gone. About one fourth of all the buildings in Tokyo went up in smoke that night anyway. More than two hundred and sixty-seven thousand buildings.

Here's what happened. We ordered three hundred and twenty-five planes to that job, and eighty-six per cent of them attacked the primary target. We lost just four-and-three-tenths per cent of all the B-29's which were airborne. Sixteen hundred and sixty-five tons of incendiary bombs went hissing down upon that city, and hot drafts from the resulting furnace tossed some of our aircraft two thousand feet above their original altitude. We burned up nearly sixteen square miles of Tokyo.

To quote General Power (later he succeeded me as commander of SAC, and was the man who led this operation in person): 'It was the greatest single disaster incurred by any enemy in military history. It was greater than the combined damage of Hiroshima and Nagasaki. There were more casualties than in any other military action in the history of the world.'

Contrary to supposition and cartoons and editorials of our enemies, I do not beam and gloat where human casualties are concerned.

I'll just quote *AAFWW II*, Volume V, page 617, and let it go at that. 'The physical destruction and loss of life at Tokyo exceeded that at Rome . . . or that of any of the great conflagrations of the western world – London, 1666 . . . Moscow, 1812 . . . Chicago, 1871 . . . San Francisco, 1906. . . . Only Japan itself, with the earthquake and fire of 1923 at Tokyo and Yokohama, had suffered so terrible a disaster. *No other air attack of the war, either in Japan or Europe, was so destructive of life and property.'*

The italics are my own.

Curtis LeMay saw the bomber's war with a general's eye: Bert Stiles was just a humble co-pilot on B-17s flying from East Anglia. He'd been a student at Colorado College till 1942 when he joined the USAAF. He arrived in Britain in March 1944, and when his bomber tour of thirty-five missions was complete, volunteered to stay in Europe and fly fighters. (He was shot down and killed in a P-51 Mustang on an escort mission to Hanover in November that year. He was just twenty-three.)

29

FROM: *Serenade to the Big Bird*
BY   Bert Stiles
     Norton, New York, 1952

Leipzig
The crews scheduled for that haul were waked up around 0300 hours. There was plenty of bitching about that.

I was so tired I felt drunk.

They told us there'd be eggs for breakfast, but there was just bacon without eggs. There was plenty of bitching about that too.

In the equipment hut I heard somebody say, 'Today I'm catching up on my sack time.'

Some other gunner said, 'I slept most of the way to Augsburg yesterday.'

Nobody said anything about the Luftwaffe. Leipzig is in there deep, but plenty of gunners bitched about taking extra ammunition. Plenty of gunners didn't take any.

Beach was flying his last mission with Langford's crew.

'We're the last of Lieutenant Newton's gang,' he said wanly.

'And I'll be the very last,' I said. 'Take it easy today.'

We had an easy ride in. I didn't feel sleepy. I just felt dazed.

There was soft fuzz over a thin solid overcast going in, but inside Germany the clouds broke up. There was haze under the cumulus and the ground showed pale green through the holes.

'We're way back,' Green said.

The group was tucked in nicely, the low squadron was up close, and Langford was doing a pretty job of flying high.

The lead and high groups of our wing looked nice too. But our group, the low, was way back and below. Our wing was the tail end, with most of the 8th up ahead.

The wing had S-ed out and called our group leader to catch up. He didn't.

If I didn't listen to the engine roar it was quiet up there. The sky was a soft sterile blue. Somehow we didn't belong there.

There was death all over the sky, the quiet threat of death, the anesthesia of cold sunlight filled the cockpit.

The lady named Death is a whore . . . Luck is a lady . . . and so is Death. . . . I don't know why. And there's no telling who they'll go for. Sometimes it's a quiet, gentle, intelligent guy. The Lady Luck strings along with him for a while, and then she hands him over to the lady named Death. Sometimes a guy comes along who can laugh in their faces. The hell with luck, and the hell with death. . . . And maybe they go for it . . . and maybe they don't.

There's no way to tell. If you could become part of the sky you might know . . . because they're always out there. The lady of Luck has a lovely face you can never quite see, and her eyes are the night itself, and her hair is probably dark and very lovely . . . but she doesn't give a damn.

And the lady named Death is sometimes lovely too, and sometimes she's a screaming horrible bitch . . . and sometimes she's a quiet one, with soft hands that rest gently on top of yours on the throttles.

The wing leader called up, 'We're starting our climb now.' We only had a half hour or so until target time.

He hadn't listened. The lead and high groups were already far above us. We were back there alone.

We never caught up after that.

'I don't like this,' Green said.

'Tuck it in,' somebody said over VHF. 'Bandits in the target area.'

I was tense and drawn taut. The sky was cold and beautifully aloof.

Green was on interphone and I was on VHF, listening for anything from the lead ship.

I heard a gun open up.

Testing, I decided.

I saw some black puffs and a couple of bright bursts.

Jesus, we're in the flak already, I thought.

Then the guns opened up. Every gun on the ship opened up. A black Focke-Wulf slid under our wing, and rolled over low.

I flipped over on interphone and fear was hot in my throat and cold in my stomach.

'Here they come.' It was Mock, I think, cool and easy, like in a church. Then his guns fired steadily.

The air was nothing but black polka dots and firecrackers from the 20-millimeters.

'Keep you eye on 'em. Keep 'em out there. . . .' It was Mock and Bossert.

'Got the one at seven . . .' Bossert or Mock. Steady.

They came through again, coming through from the tail.

I saw two Forts blow up out at four o'clock. Some other group.

A trio of gray ones whipped past under the wing and rolled away at two o'clock. Black crosses on gray wings . . . 109s.

A night-fighter Focke-Wulf moved up almost in formation with us, right outside the window, throwing shells into somebody up ahead. Somebody powdered him.

One came around at ten o'clock . . . and the nose guns opened up on him. He rolled over and fell away . . . maybe there was smoke. . . .

The instruments were fine. Green looked okay. My breath was in short gasps.

'Better give me everything,' Green said. Steady voice.

I jacked-up the RPM up to the hilt.

They were queuing up again back at four and six and eight. A hundred of them . . . maybe two hundred . . . getting set to come through again . . . fifteen or twenty abreast. . . .

. . . I looked up at the other wing-ship. The whole stabilizer was gone. I could see blue sky through there . . . but the rudder still worked . . . still flapped . . . then his wing flared up . . . he fell off to the right.

We were flying off Langford, but he was gone . . . sagging off low at three o'clock. Green slid us in under the lead squadron. Langford was in a dive . . . four or five planes were after him . . . coming in . . . letting them have it . . . swinging out . . . and coming in again. . . . Beach was in that ball . . . poor goddamn Beach. . . .

'Here they come!'

'Four o'clock level.'

'Take that one at six.'

All the guns were going again.

There wasn't any hope at all . . . just waiting for it . . . just sitting there hunched up . . . jerking around to check the right side . . . jerking back to check the instruments . . . everything okay . . . just waiting for it. . . .

They came through six times, I guess . . . maybe five . . . maybe seven . . . queuing up back there . . . coming in . . . throwing those 20s in there.

. . . we were hit . . .

. . . the whole low squadron was gone . . . blown up . . . burned up . . . shot to hell . . . one guy got out of that.

. . . we were the only ones left in the high . . . tucked in under the lead. The lead squadron was okay . . . we snuggled up almost under the tail guns. They were firing steadily . . . the shell cases were dropping down and going through the cowling . . . smashing against the plexiglass . . . chipping away at the windshields . . . coming steady . . . coming all the time . . . then his guns must have burned out . . .

. . . there were a few 51s back there . . . four against a hundred . . . maybe eight. . . .

'Don't shoot that 51,' Mock again, cool as hell . . .

I punched the wheel forward. A burning plane was nosing over us.

Green nodded, kept on flying. . . .

The guns were going . . . not all of them any more . . . some of them were out . . . burned out . . . maybe.

And then it was over. They went away.

We closed up and dropped our bombs.

Six out of twelve gone.

We turned off the target, waiting for them . . . knowing they'd be back . . . cold . . . waiting for them. . . .

There was a flow to it . . . we were moving . . . we were always moving . . . sliding along through the dead sky. . . .

I flicked back to VHF.

No bandits called off.

Then, I heard, '. . . is my wing on fire? . . . will you check to see

187

if my wing is on fire? . . .'

He gave his call sign. It was the lead ship.

We were right underneath. We pulled up even closer.

'You're okay,' I broke the safety wire on the transmitter. 'You're okay . . . baby . . . your wing is okay. No smoke . . . no flame . . . stay in there, baby.'

It was more of a prayer.

'. . . I'm bailing out my crew. . . .'

I couldn't see any flame. I wasn't sure it was the same plane. But they were pulling out to the side.

All my buddies. Maurie . . . Uggie . . .

I told Green. 'We better get back to the main group . . . we better get back there fast . . .'

We banked over. I saw the rear door come off and flip away end over end in the slipstream. Then the front door, then something else. . . . maybe a guy doing a delayed jump. It didn't look like a guy very much.

It must have been set up on automatic pilot. It flew along out there for half an hour. If they jumped they were delayed jumps.

Maybe they made it.

We found a place under the wing lead.

I reached over and touched Green. What a guy. Then I felt the control column. Good airplane . . . still flying . . . still living. . . .

Everybody was talking.

Nobody knew what anybody said.

There was a sort of beautiful dazed wonder in the air . . .

. . . still here. . . . Still living . . . still breathing.

And then it came through . . . the thought of all those guys . . . those good guys . . . cooked and smashed and down there somewhere, dead or chopped up or headed for some Stalag.

We were never in that formation. We were all alone, trailing low.

From the day you first get in a 17 they say formation flying is the secret.

They tell you over and over. Keep those planes tucked in and you'll come home.

The ride home was easy. They never came back.

The sky was a soft unbelievable blue. The land was green, never so green.

When we got away from the Continent we began to come apart. Green took off his mask.

There weren't any words, but we tried to say them.

'Jesus, you're here,' I said.

'I'm awfully proud of them,' he said quietly.

Bradley came down out of the turret. His face was nothing but teeth. I mussed up his hair, and he beat on me.

The interphone was jammed.

'. . . all I could do was pray . . . and keep praying.' McAvoy had to stay in the radio room the whole time, seeing nothing, doing nothing. . . .

'You can be the chaplain,' Mock said. His voice just the same, only he was laughing a little now.

'. . . if they say go tomorrow . . . I'll hand in my wings. . . . I'll hand in every other goddamn thing . . . but I won't fly tomorrow . . .' Tolbert was positive.

. . . if Langford went down . . . that meant Fletch . . . Fletch and Johnny O'Leary and Beach . . .

. . . and all the others . . . Maurie had long black eyelashes, and sort of Persian's eyes . . . sort of the walking symbol of sex . . . and what a guy . . . maybe he made it . . . maybe he got out. . . .

It was low tide. The clouds were under us again, almost solid, and then I saw a beach through a hole . . . white sand and England.

There was never anywhere as beautiful as that.

We were home.

Green made a sweet landing. We opened up the side windows and looked around. Everything looked different. There was too much light, too much green . . . just too much . . .

We were home. . . .

They sent us out to get knocked off and we came home.

And then we taxied past E-East.

'Jesus, that's Langford,' I grabbed Green.

It was. Even from there we could see they were shot to hell. Their tail was all shot up . . . one wing was ripped and chopped away.

Green swung around into place, and I cut the engines.

. . . we were home . . .

There were empty spaces where ships were supposed to be,

189

where they'd be again in a day, as soon as ATC could fly them down.

We started to talk to people. There were all kinds of people. Jerry, a crew chief, came up and asked us about the guys on the other wing. We told him. Blown up.

. . . honest to God . . . we were really home. . . .

The 20-millimeter hit our wing . . . blew up inside . . . blew away part of the top of number two gas tank . . . blew hell out of everything inside there . . . puffed out the leading edge . . . blew out an inspection panel.

We didn't even lose any gas.

We didn't even blow up.

I stood back by the tail and looked at the hole. I could feel the ground, and I wanted to take my shoes off. Every time I breathed, I knew it.

I could look out into the sky over the hangar and say thank you to the lady of the luck. She stayed.

I was all ripped apart. Part of me was dead, and part of me was wild, ready to take off, and part of me was just shaky and twisted and useless.

Maybe I told it a thousand times.

I could listen to myself. I could talk, and start my voice going, and step back and listen to it.

I went down to Thompson's room, and he listened. He listened a couple of times.

It was a pretty quiet place. Eight ships out of a group is a quiet day at any base.

Colonel Terry just got married. Thompson didn't know about it. I went back to my room and sat across the room from Langford and kept telling myself it was him.

'When I saw you there were at least eight of them,' I said. 'Just coming in, and pulling out, and coming in.' I showed him with my hands.

Then Fletch came in.

Then I thought about Beach.

Beach got three at least. He shot up every shell he had, and got three.

He came over after interrogation.

'I guess they can't kill us Denver guys,' he said. He didn't believe it either. He was all through.

'Jesus,' I said, 'I sure thought they had you.'

Green came in with O'Leary.

'I knew you were down,' O'Leary said. 'I told everyone.'

Green smiled. He looked okay, 'We're on pass,' he said quietly. 'Let's get out of here.'

I wanted to touch him again. I wanted to tell him I was glad I was on his crew, and it was the best goddamn crew I'd ever heard of, but I didn't say anything, and he didn't either.

I got out my typewriter and started a letter to my folks.

And then it came in again . . . all those guys . . . all those good guys . . . shot to hell . . . or captured . . . or hiding there waiting for it.

. . . waiting for it. . . .

Then I came all apart, and cried like a little kid. . . . I could watch myself, and hear myself, and I couldn't do a goddamn thing.

. . . just pieces of a guy . . . pieces of bertstiles all over the room . . . maybe some of the pieces were still over there.

And then it was all right. I went in and washed my face. Green was calling up about trains, standing there in his shorts.

'I think the boys need a rest,' he said. 'You going in?'

'I'll meet you in London at high noon,' I said. 'Lobby of the Regent Palace.'

'Okay,' he said. 'Get a good night's sleep.'

'Meet you there,' I said.

But I didn't.

They sent me to the Flak House. There was an opening, and the squadron sent me.

'What odd things most of us do when under stress,' wrote Pappy
Boyington, and he ended his autobiography: 'Just name a hero and
I'll prove he's a bum!' Boyington was a considerable World War
Two hero determined to prove himself a bum: at the end of his book,
he is a post-war drunkard, drifting from job to job till rescued by
Alcoholics Anonymous.

Boyington was no stranger to stress: he had a long war, starting
with Chennault's Flying Tigers in China, and finishing as a severely
wounded prisoner of the Japanese. He had scored six kills in China,
and added twenty-two more with the US Marines – though he gave
the USMC chiefs almost as much trouble as he did the Japanese. He
was, said one biographer, 'Unsurpassed by any other American
fighter pilot for sheer colour, flamboyance and fighting skill . . . in his
own way, "Pappy" got the job done, even if his plain speaking,
frankness and insobriety made him seem less than a gentleman . . .'
As a fighter pilot he was superb, as strong as an ox (he'd been a
wrestler in college), a marvellous shot, and revered by his men – a
'gang of outcast, rambunctious pilots he led as the "Black Sheep"
squadron'. Together they shot down more Japanese aircraft than any
other squadron and that, after all, was their task.

His reluctant-hero's tale Baa Baa Black Sheep was first a best-
seller, then long out of print, then suddenly the basis for a television
series in the USA. In the excerpt from it that follows, he describes his
last mission with his squadron, VMF-214, and the pressures upon
him that led up to it.

30
FROM: *Baa Baa Black Sheep*
BY 'Pappy' Boyington
Putnam's, New York, 1958

During one of these daily hops over Rabaul I had
reached a definite climax in my flying career without too much
effort. I shot down my twenty-fifth plane on December 27. And if
I thought that I ever had any troubles previously, they were a
drop in the bucket to what followed.

There was nothing at all spectacular in this single victory, but it
so happened that this left me just one short of the record jointly

held by Eddie Rickenbacker of World War I and Joe Foss of World War II. Then everybody, it seemed to me, clamored for me to break the two-way tie. The reason for all the anxiety was caused by my having only ten more days to accomplish it; 214 was very near completion of its third tour, and everyone knew I would never have another chance. My combat-pilot days would close in ten days, win, lose, or draw.

Everyone was lending a hand, it seemed, but I sort of figured there was too much help. Anyway, I showed my appreciation by putting everything I had left into my final efforts. I started flying afternoons as well as mornings, and in bad weather in addition to good weather.

One pre-dawn take-off was in absolute zero-zero conditions, and all we had for references were two large searchlights on the end of the Vella strip, one aimed vertically and the other horizontally. I wasn't questioned by the tower whether I had an instrument ticket, like I am today. My last words to my pilots before I started my take-off through the fog were: 'Please listen to me, fellows, and have complete faith in your instruments. If you dare to take one look out of the cockpit after you pass the searchlights, you're dead.'

As I had always been unaccustomed to help or encouragement in the past, all the extra help did nothing more than upset me. But I couldn't have slowed down or stopped if I had wanted to, simply because nobody would let me.

One fight made me desperate when I could not see to shoot with accuracy, because I wasn't able to see well enough through the oil-smeared windshield. After several fruitless attempts I pulled off to one side of the fight and tried to do something to correct it. I unbuckled my safety belt and climbed from my parachute harness, then opened the hood and stood up against the slip-stream, trying my darnedest to wipe off the oil with my handkerchief. It was no use; the oil leak made it impossible for me to aim with any better accuracy than someone who had left his glasses at home.

Soon I began to believe that I was jinxed. Twice I returned with bullet holes in my plane as my only reward. Twice I ran into a souped-up version of the Zero known as the Tojo. Though not quite as maneuverable as the original, it was considerably faster and had a greater rate of climb. Still no shoot-downs, and I was lucky the Nips didn't get me instead.

Doc Ream was really concerned over the way I was affected by the pressure, suggesting we call a halt to the whole affair. He said that there were plenty of medical reasons for calling all bets off. But I knew I couldn't stop. Whether I died in the attempt made no difference. Anyway, my last combat tour would be up in a few days, and I would be shipped back to the United States. I said: 'Thanks for the out, Doc, but I guess I better go for broke, as the Hawaiians say.'

Never had I felt as tired and dejected as I did when I flew into Vella one afternoon in late December. Another futile attempt was behind me. The bullet holes in my plane were a far cry from the record I was striving to bring back. I was dead tired, I had counted upon the day ending, but a pilot had crawled up on my wing after I had cut my engine, and he had something important to say.

Marion Carl was scheduled to take several flights that afternoon to Bougainville, where they were to remain overnight, taking off on the following morning for a sweep. He said: 'Greg, I want to give you a chance to break the record. You take my flight because you're so close I think you are entitled to it. I've got seventeen, but I still have loads of time left, and you haven't.'

Carl had been out previously in the Guadalcanal days as a captain, piling up a number of planes to his credit, and was then back for the second time, as a squadron commander. He had just been promoted to major, and it was true that many chances were coming up for him. Great person that Marion Carl is, he was trying to give a tired old pilot a last crack at the title, even though it was at his own expense.

I can never forget George Ashmun's thin, pale face when I mentioned where I was going, and he insisted that he go along as my wingman. Maybe George knew that I was going to have to take little particles of tobacco from a cigarette, placing them into the corners of my eyes to make them smart so that I'd stay awake.

Those close to me were conscious of what kind of shape I was in, and they were honestly concerned. But I was also happy to find others I hadn't thought of at the time who were concerned for

194

my welfare as well, though in most cases I didn't discover this until after the war. And that was by mail.

Some of the letters were clever, but I especially remember one from a chap who I imagine must have been about eighteen. He wrote me that, after I was missing in action, his partner, 'Grease Neck', who worked on a plane with him, had said that I was gone for good, and the first chap said : 'I bet you he isn't.' The outcome of the discussion was that each bet a hundred and fifty dollars, one that I would, the other that I would not, be back home six months after the war was over. The six-months business referred to the fact that if you are missing six months after the war is over you are officially declared dead. And at one time I had said, just as a morale builder to the other pilots so that they would not worry about me : 'Don't worry if I'm ever missing, because I'll see you in Dago and we'll throw a party six months after the war is over.'

I had said it by coincidence just before taking off on what turned out to be my last fight, but the words apparently had stuck in their minds.

But to get back to this letter from the young chap, he told me how thrilled he was about my being home, and he told me about this bet he had made with 'Grease Neck', and how he had just collected that hundred and fifty, and that he was going to spend the entire amount on highballs in my honor in San Diego.

It was a great feeling to get those letters and know that the boys really wanted to see you home – bets or no bets. I also hope, because I never heard any more from this young fellow, that he didn't end up in the local bastille while celebrating in my honor.

My thoughts then are much the same now in many respects. Championships in anything must be a weird institution. So often there is but a hairline difference between the champion and the runner-up. This must go for boxing and tennis, football and baseball. In my case it was something else, the record for the number of planes shot down by a United States flyer, and I was still having quite a time trying to break it.

After getting twenty-five planes, most of them on missions two hundred miles or better into enemy air, I had gone out day after day, had had many a nice opportunity, but always fate seemed to step in and cheat me : the times there was oil on the windshield and I couldn't see any of the planes I fired into go down or flame ;

195

the times my plane was shot up. Nothing seemed to work for me. Then everybody, including the pressmen, kept crowding me and asking: 'Go ahead; when are you going to beat the record?' I was practically nuts.

Then came the day when the record finally was broken, but, as so often happens with one in life, it was broken without much of a gallery. And in this case without even a return.

It was before dawn on January 3, 1944, on Bougainville. I was having baked beans for breakfast at the edge of the airstrip the Seabees had built, after the Marines had taken a small chunk of land on the beach. As I ate the beans, I glanced over at row after row of white crosses, too far away and too dark to read the names. But I didn't have to. I knew that each cross marked the final resting place of some Marine who had gone as far as he was able in this mortal world of ours.

Before taking off everything seemed to be wrong that morning. My plane wasn't ready and I had to switch to another. At the last minute the ground crew got my original plane in order and I scampered back into that. I was to lead a fighter sweep over Rabaul, meaning two hundred miles over enemy waters and territory again.

We coasted over at about twenty thousand feet to Rabaul. A few hazy clouds and cloud banks were hanging around – not much different from a lot of other days.

The fellow flying my wing was Captain George Ashmun, New York City. He had told me before the mission: 'You go ahead and shoot all you want, Gramps. All I'll do is keep them off your tail.'

This boy was another who wanted me to beat that record, and was offering to stick his neck way out in the bargain.

I spotted a few planes coming up through the loosely scattered clouds and signaled to the pilots in back of me: 'Go down and get to work.'

George and I dove first. I poured a long burst into the first enemy plane that approached, and a fraction of a second later saw the Nip pilot catapult out and the plane itself break out into fire.

George screamed over the radio: 'Gramps, you got a flamer!'

Then he and I went down lower into the fight after the rest of the enemy planes. We figured that the whole pack of our planes was going to follow us down, but the clouds must have obscured us from their view. Anyway, George and I were not paying too much attention, just figuring that the rest of the boys would be with us in a few seconds, as usually was the case.

Finding approximately ten enemy planes, George and I commenced firing. What we saw coming from above we thought were our own planes – but they were not. We were being jumped by about twenty planes.

George and I scissored in the conventional Thatch-weave way, protecting each other's blank spots, the rear ends of our fighters. In doing this I saw George shoot a burst into a plane and it turned away from us, plunging downward, all on fire. A second later I did the same to another plane. But it was then that I saw George's plane start to throw smoke, and down he went in a half glide. I sensed something was horribly wrong with him. I screamed at him: 'For God's sake, George, dive!'

Our planes could dive away from practically anything the Nips had out there at the time, except perhaps a Tony. But apparently George never heard me or could do nothing about it if he had. He just kept going down in a half glide.

Time and time again I screamed at him: 'For God's sake, George, dive straight down!' But he didn't even flutter an aileron in answer to me.

I climbed in behind the Nip planes that were plugging at him on the way down to the water. There were so many of them I wasn't even bothering to use my electric gun sight consciously, but continued to seesaw back and forth on my rudder pedals, trying to spray them all in general, trying to get them off George to give him a chance to bail out or dive – or do something at least.

But the same thing that was happening to him was now happening to me. I could feel the impact of the enemy fire against my armor plate, behind my back, like hail on a tin roof. I could see enemy shots progressing along my wing tips, making patterns.

George's plane burst into flames and a moment later crashed into the water. At that point there was nothing left for me to do. I had done everything I could. I decided to get the hell away from the Nips. I threw everything in the cockpit all the way forward – this means full speed ahead – and nosed my plane over to pick up extra speed until I was forced by the water to level off. I had gone practically a half mile at a speed of about four hundred knots, when all of a sudden my main gas tank went up in flames in front of my very eyes. The sensation was much the same as opening the door of a furnace and sticking one's head into the thing.

Though I was about a hundred feet off the water, I didn't have a chance of trying to gain altitude. I was fully aware that if I tried to gain altitude for a bail-out I would be fried in a few more seconds.

At first, being kind of stunned, I thought: 'Well, you finally got it, didn't you, wise guy?' and then I thought: 'Oh, no you didn't!' There was only one thing left to do. I reached for the rip cord with my right hand and released the safety belt with my left, putting both feet on the stick and kicking it all the way forward with all my strength. My body was given centrifugal force when I kicked the stick in this manner. My body for an instant weighed well over a ton, I imagine. If I had had a third hand I could have opened the canopy. But all I could do was to give myself this propulsion. It either jettisoned me right up through the canopy or tore the canopy off. I don't know which.

There was a jerk that snapped my head and I knew my chute had caught – what a relief. Then I felt an awful slam on my side – no time to pendulum – just boom-boom and I was in the water.

The cool water around my face sort of took the stunned sensation away from my head. Looking up, I could see a flight of four Japanese Zeros. They had started a game of tag with me in the water. And by playing tag, I mean they began taking turns strafing me.

I started diving, making soundings in the old St George Channel. At first I could dive about six feet, but this lessened to four, and gradually I lost so much of my strength that, when the Zeros made their strafing runs at me, I could just barely duck my head under the water. I think they ran out of ammunition, for after a while they left me. Or my efforts in the water became so feeble that maybe they figured they had killed me.

The best thing to do, I thought, was to tread water until nightfall. I had a little package with a rubber raft in it. But I didn't want to take a chance on opening it for fear they might go back to Rabaul, rearm, and return to strafe the raft. Then I would have been a goner for certain.

I was having such a difficult time treading water, getting weaker and weaker, that I realized something else would have to be done real quickly. My 'Mae West' wouldn't work at all, so I shed all my clothes while I was treading away; shoes, fatigues, and everything else. But after two hours of this I knew that I couldn't keep it up any longer. It would have to be the life raft or nothing. And if the life raft didn't work – if it too should prove all shot full of holes – then I decided: 'It's au revoir. That's all there is to it.'

I pulled the cord on the raft, the cord that released the bottle of compressed air, and the little raft popped right up and filled. I was able to climb aboard, and after getting aboard I started looking around, sort of taking inventory.

I looked at my Mae West. If the Nips came back and strafed me again, I wanted to be darned sure that it would be in working order. If I had that, I could dive around under the water while they were strafing me, and would not need the raft. I had noticed some tears in the jacket, which I fully intended to get busy and patch up, but the patching equipment that came with the raft contained patches for about twenty-five holes.

'It would be better first, though,' I decided, 'to count the holes in this darned jacket.' I counted, and there were more than two hundred.

'I'm going to save these patches for something better than this.' With that I tossed the jacket overboard to the fish. It was of no use to me.

Then for the first time – and this may seem strange – I noticed that I was wounded, not just a little bit, but a whole lot. I hadn't noticed it while in the water, but here in the raft I certainly noticed it now. Pieces of my scalp, with hair on the pieces, were hanging down in front of my face.

My left ear was almost torn off. My arms and shoulders contained holes and shrapnel. I looked at my legs. My left ankle was shattered from a twenty-millimeter-cannon shot. The calf of my left leg had, I surmised, a 7·7 bullet through it. In my groin I had been shot completely through the leg by twenty-millimeter shrapnel. Inside of my leg was a gash bigger than my fist.

'I'll get out my first-aid equipment from my jungle pack. I'd better start patching this stuff up.'

I kept talking to myself like that. I had lots of time. The Pacific would wait.

Even to my watch, which was smashed, I talked also. The

impact had crushed it at a quarter to eight on the early morning raid. But I said to it: 'I'll have a nice long day to fix you up.'

I didn't, though. Instead, I spent about two hours trying to bandage myself. It was difficult getting out these bandages, for the waves that day in the old South Pacific were about seven feet or so long. They are hard enough to ride on a comparatively calm day, and the day wasn't calm.

After I had bandaged myself as well as I could, I started looking around to see if I could tell where I was or where I was drifting. I found that my raft contained only one paddle instead of the customary two. So this one little paddle, which fitted over the hand much like an odd sort of glove, was not of much use to me.

Talking to myself, I said: 'This is like being up shit creek without a paddle.'

Far off to the south, as I drifted, I could see the distant shore of New Britain. Far to the north were the shores of New Ireland. Maybe in time I could have made one or the other of these islands. I don't know. But there is something odd about drifting that I may as well record. All of us have read, or have been told, the thoughts that have gone through other men under similar circumstances. But in my case it was a little tune that Moon Mullin had originated. And now it kept going through my mind, bothering me, and I couldn't forget it. It was always there, running on and on:

On a rowboat at Rabaul,
    On a rowboat at Rabaul . . .

The waves continued singing it to me as they slapped my rubber boat. It could have been much the same, perhaps, as when riding on a train, and the rails and the wheels clicking away, pounding out some tune, over and over, and never stopping.

The waves against this little rubber boat, against the bottom of it, against the sides of it, continued pounding out:

. . . On a rowboat at Rabaul,
    You're not behind a plow . . .

And I thought: 'Oh, Moon Mullin, if only I had you here, I'd wring your doggone neck for ever composing the damn song.'

200

What might it have been like, towards the end, if we, the Allies, had lost the war? Full of fighting as bitter, desperate and futile as another German fighter pilot describes here, no doubt. And if all is fair in war, airmen have somehow never accepted shooting at an enemy as he parachutes to safety to be quite cricket – illogical, because if you let him go today, he could be back in the air and shooting you down tomorrow. It seems that shooting at parachutes was not just a dastardly Hun trick, as we were taught: some of us did it too.

31

FROM: *I Flew for the Führer*
BY   Heinz Knoke
       Evans, London, 1953

28th August, 1944.
The enemy try to cross the Seine on pontoon bridges between Vernon and Mantes. Ceaseless fighter patrols form an umbrella, together with a cordon of concentrated flak to protect the crossing.

During the six missions in this sector yesterday the Squadron lost twelve aircraft. We are finished.

This morning the Squadron serviceability report lists only four aircraft as operational. Two others with badly twisted fuselages are capable of non-operational flying only. They are such battered old crates that I am not going to be responsible for sending any of my men into combat in them.

So at 0600 hours there is a telephone call from the Chief Staff Officers at Corps Headquarters. He gives me a furious reprimand.

'This morning you reported only four aircraft available for operations. I have just learned that you can still fly six. Are you crazy? Do you realize the seriousness of the situation? It is nothing but sabotage; and I am not going to tolerate it. Every one of your aircraft is to fly. That is an order!'

He is bellowing like a bull. I have never been reprimanded like this since I finished my basic training as an Air Force recruit. I am so furious that I can hardly control my rage. Why should I have to listen to that arrogant ape? He even has the nerve to accuse *me* of sabotage! Chairborne strategists and heroes of the staff make me sick. They know nothing of the problems at the Front which we are up against, and they care even less.

I decide to fly one of the worn-out crates myself, and let my wingman, Corporal Döring, take the other. According to the operation orders, we are to take off at 0800 hours and rendezvous with the other Squadrons of the Wing over Soissons. I am then to take over command of the entire fighter formation.

Two minutes before zero hour the engines are started. We roll out from the camouflaged bushes and turn into wind. There is no runway, only a length of soft field. My aircraft lumbers along, gathering flying speed with difficulty, and it is all I can do to coax the old crate into the air in time to clear the trees at the far end of the field.

Döring tries to climb too soon and stalls. His left wing drops and he plunges to crash into the trees. Flames belch forth. Döring is instantly killed – and then we are five.

The order from the Chief Staff Officer at Corps is worse than insanity; it amounts to nothing less than murder!

Base reports by radio to advise me that the other Squadrons are unable to leave the ground, because their fields are being strafed by enemy fighter-bombers.

'Go to sector Siegfried-Gustav.'

North of Soissons lies the little town of Tergnier. It is a large railway junction at the point where the Somme Canal meets the River Oise. As a conspicuous landmark it is visible from a great distance. Above it, the Third Squadron of No. 1 Fighter Wing now fights its last air battle against the Americans over French territory.

We encounter more than sixty Thunderbolts and Mustangs in this area. There can be no escape: it is the end. All that remains is for me to give the order to attack. Thus at least a moral victory can still be claimed by my men and myself.

Base still try to give me orders. I turn off the radio; to hell with them now!

My aircraft cannot climb above 10,000 feet. It is very slow and unresponsive. I feel certain that this is its last flight.

The battle does not last for more than a few minutes. Corporal Wagner is the first to be shot down; he does not escape from his flaming aircraft.

Then I see another aircraft on fire, and Flight Sergeant Freigang bales out. His wingman goes down in flames a few moments later.

That leaves only my wingman, Sergeant Ickes, and myself. For us there can be no way out. If this is to be the end, I can only sell my life as dearly as possible. If I ram one of the Yanks I shall be able to take him with me. . . .

Tracers converge on us from all sides. Bullets slam my aircraft like hailstones, and it gradually loses forward speed. Ickes remains close beside me. I keep on circling in as tight a turn as possible. A Mustang gets on to my tail. I am unable to shake it off. My plane is too sluggish, as if it felt too tired to fly any farther. More bullets come slamming into the fuselage behind my head.

With a last burst of power from the engine I pull the aircraft up in a climb, half-roll on to the side, and cut the throttle. The Mustang on my tail has not anticipated this. It shoots past, and now it is in front of me and a little below. I distinctly see the face of the pilot as he turns his head to look for me. Too late, he attempts to escape by diving. I am on him now. I can at least ram him if I cannot shoot him down. I feel icy cold. My only emotion, for the first time in life, is intense personal hatred of my enemy; my only desire is to destroy him.

The gap closes rapidly : we are only a few feet apart. My salvoes slam into the fuselage : I am aiming for the pilot. His engine bursts into flames. We shall go down together !

There is a violent jolt at the first impact. I see my right wing fold and break away. In a split-second I jettison the canopy and am out of the seat. There is a fierce blast of flame as I am thrown clear, while Messerschmitt and Thunderbolt are fused together in a single ball of fire.

A few moments later my parachute mushrooms overhead. Six to eight hundred feet away and a little higher up there is another open parachute. Ickes.

Overhead and all around us the Americans continue to thunder, circling and milling around like mad. It is a few minutes before it dawns on them that by this time not a single Messerschmitt is left in the sky.

A Thunderbolt comes diving towards me. It opens fire ! For age-long seconds my heart stops beating. I throw up my hands and cover my face. . . .

Missed!

Round it comes again, this time firing at Ickes. I can only watch, while the body of my comrade suddenly slumps lifeless. Poor Ickes!

What a foul and dirty way of fighting! War is not the same as a football match; but there is still such a thing as fair play.

I come down to land in a forest clearing. I have no idea whether I am on the German side or behind the enemy lines. Therefore, I start by hiding in the dense undergrowth.

Overhead, the Americans fly away to the west.

It is wonderful to be able to relax. I light a cigarette and lie back on the parachute shrouds, gratefully inhaling the soothing smoke.

As a precautionary measure, I remove the rank-badges from my shoulders and stuff the German Gold Cross into my pocket.

I happen to be wearing an American leather flying-jacket, a dark blue silk sports shirt, rather faded trousers, and black walking-shoes. The whole effect is so un-Prussian that no one will recognize me immediately as a German.

My caution very soon proves justified.

About fifteen minutes after my descent I notice four French civilians at the other end of the clearing. They gesticulate wildly as they talk. With my school French I am able to understand that they must be looking for me. Each one of them wants to search in a different place. I gather that they are under the impression that the parachutist is an American. All four carry arms. Evidently they are underground terrorists of the French Resistance.

I grope for the pistol hidden under my bulky leather jacket.

The four start combing the bushy undergrowth. Discovery is inevitable, sooner or later, so I decide to go out to meet them.

All four look surprised to see me. Four tommy-guns swing round to cover me. Now is the time for me to keep calm and clever. The French have a bitter hatred for us Germans, from the bottom of their temperamental souls. Not that I blame them; no doubt I should feel the same in their place. But if the bastards ever guess that I am German it will mean a bellyful of lead, as sure as God made little apples.

So I walk up coolly, and in the friendliest possible manner greet them in English: 'Hello, boys!'

The stern faces of the bandits gradually relax into smiles. They take me for a Yank.

In my best American accent I then proceed to ask them in very broken French to help me find my 'comrades':–

'*Voulez-vous aider moi trouver mes camarades?*'

They immediately explain the position to me. An American armoured unit with Sherman tanks is a little over a mile away. We must be very cautious, however, as the place is still swarming with the lousy Boches. Fighting is going on all round us. In fact I can now hear for the first time the distant gunfire.

The tallest of the Frenchmen – a thoroughly repulsive-looking type – carries a German tommy-gun. I do not like the look of him at all. He remains in the background, suspiciously quiet. Does he doubt that I am what I appear to be?

We make our way through the dense forest until we come to a railway embankment.

There is a sudden chattering from a German machine-gun; it sounds like an M.G. 45, and is quite close. The three Frenchies in front drop flat on their faces. The tall bandit remains standing close to me; evidently he is not going to let me out of his sight. From the other side of the railway comes the grinding clatter of heavy tank engines.

I ask where the line goes.

'*Vers Amiens.*'

Amiens!!? Did I really drift so far west during the dogfight? The city has been in American hands for some time. Blast! I have no desire to spend the rest of the war sitting in a prison camp somewhere in the U.S.A.

The nearest town, I learn, is Nesle. Then the Somme Canal must be somewhere to the north. According to the early morning intelligence reports, it is still being held by our forces. I shall have to head north. But how am I going to get rid of those blasted Frenchies?

More heavy gun-fire is heard. The sound comes from the west. The Frenchmen cautiously cross the tracks and wave to me to follow them. The big fellow stays on my tail with his tommy-gun, otherwise I would be able to bolt back into the forest and then make my way round to the north.

A few hundred yards farther on we come to a highway. It cuts across the landscape, straight as an arrow, and is visible for miles.

The chattering of several machine-guns is again heard off to the left. The first three Frenchmen cautiously cross the road. The big fellow takes two or three paces after them, then turns towards me. Our eyes meet. I can tell that he recognizes me. I must be off! There will be no second chance to break away; it must be now or never.

I dash back towards the forest. Then the big fellow is coming after me, before his comrades realize what is happening. He lifts his tommy-gun and starts firing. I drop behind a bank of earth. Bullets thud into the ground all round me.

The bandit empties his clip. He must take his eyes off the target long enough to insert a new one. There is just enough time to draw my pistol and snap up the safety-catch. I leap at the big fellow, who is raising his tommy-gun again, and fire once. It is enough. He goes down with a bullet in the head.

I take his tommy-gun. 'Sorry, my friend, but he who hits first lives longest.'

Panting, I struggle through the dense undergrowth. Twigs and branches lash at my face. The other three Frenchmen are left behind.

Fifteen minutes later I run into a German patrol. They are soldiers from an armoured unit.

The Mary Celeste *was that fatefully famous ship found sailing empty and unmanned, her crew all vanished without trace, and without trace of any explanation of what might have been their fate, or why. But the* Mary Celeste *was not unique. All down history and continuing into modern times there have been instances of ships' crews vanishing like ghosts at daybreak. In fact, aviation has its own* Mary Celeste.

32

The Blimp that Came Home – Without its Crew

By Jack Pearl

QUOTED IN: Lighter-than-Air Flight
EDITED BY Lt. Col. C. V. Glines, USAF
Franklin Watts, New York, 1965

It was the summer of 1942. World War II was slightly more than eight months old. Rommel's Afrika Korps was driving through Egypt toward Alexandria, and in Russia, the Nazis had surrounded Stalingrad. In the Pacific, the United States had still not recovered from Pearl Harbor. The Japanese had overrun most of Asia and were still threatening Australia – seemingly undaunted by the Marines' invasion of Guadalcanal. To make matters worse, our long supply lines to the Pacific and England were in serious danger of being choked off by an intensive Axis submarine campaign – both in the Atlantic and Pacific Oceans.

At that time, the most effective antisubmarine weapon developed by the United States Navy was the blimp, or 'rubber cow'.

The blimp – a small-scale version of the dirigibles or Zeppelins so popular in the 1930s – was ideal for convoy escort duty. A submarine lurking on the flanks or tail of a convoy leaves only a subtle spoor: a brief froth of bubbles after a periscope descends, a quickly dissipated slick of oil, a vague shadow in the depths – all of these telltale clues camouflaged by millions of whitecaps and the flash of sunlight on the surface of the sea.

It is only by the slimmest chance that any one or all of these signs can be spotted from the heaving deck of a destroyer, or from a high-speed airplane, under which the sea heaves in a confused blur. The blimp, on the other hand, can hover motionless above a suspicious sign in the water and study it at leisure. If the observers decide the shadow or slick marks the location of an enemy sub, they can radio its precise location to a warship; or they can attack it themselves with depth charges or small bombs slung to the undercarriage of the cabin.

The blimp was particularly valuable for convoy work off the Pacific coastline, where the sea is frequently shrouded in fog, and overcast a good part of each day – rendering conventional air cover virtually useless. Since Navy blimps were a common wartime sight along the West Coast, the residents of Daly City, California, a southern suburb of San Francisco, scarcely noticed when one came cruising in from the Pacific at an altitude of about one thousand feet at 10.45 a.m. on August 16, 1942.

Ten minutes later, however, all of Daly City noticed, when it became apparent that the blimp – with the big letters *L-8* stenciled on its gasbag – was about to attempt a landing on the main street of the town. All traffic stopped, local housewives crowded together at open windows, fire engines came speeding to the scene with sirens wailing, and the Daly City police chief was itching to get his hands on those 'crazy Navy barnstormers'.

At exactly 11.00 a.m. the blimp settled to the ground gently on the single landing wheel beneath its cabin and its small auxiliary tail-fin wheel. Such a 'taxi landing' was considered a stylish maneuver for a blimp pilot on an airfield. In a narrow city street, it was a spectacular feat. The sole damage to the ship was a couple of bent propellers. But one thing puzzled the spectators who saw the *L-8* land. Peering through the sweeping cabin windows, they could not spot either of the ship's two-man crew! Were they hiding? Were they injured and lying out of sight on the floor? Police and firemen converged on the airship, secured her, then opened the cabin door.

'My God!' said the first man to climb inside. 'There's nobody here! The damned thing landed by itself!'

It was true. The *L-8* was a derelict. Her crewmen had vanished. Unbelieving firemen slashed open the gasbag, thinking that someone might be hiding in one of its compartments. But the two-hundred-and-fifty-foot long airship was as empty as that famous phantom of the sea, the *Mary Celeste*.

'There was something eerie about it,' one of the firemen later told a reporter. 'We got chills down our spines, and we couldn't wait till we got out of there.'

Within the hour, a naval salvage party arrived from Moffett Field, the blimp's base. The technicians who inspected the *L-8* from stem to stern were infected by the same uneasiness as the townspeople. There was no logical or apparent reason why the crewmen should have abandoned the ship. There was plenty of fuel in the tanks; the fuel lines to the engines were clear and open; the engines were in operating order; the ignition switches were on; and the two throttles were open, one full, one half, as if the blimp had been turning. There was no indication of fire or of any other emergency.

Why had the crew of the *L-8* abandoned her?

Then there were the mysteries with the mystery. The blimp's radio was in good working order, although the *L-8* had abruptly broken off communications with Moffett Field at 8.00 a.m. Why? It was logical that the rubber life jackets were missing, because crewmen were required to wear them at all times on convoy patrol. But the life raft was still securely in place. Why? If the crew had voluntarily abandoned the blimp over the ocean, they almost certainly would have taken along the life raft with its survival stores and flares.

And strangest of all, the confidential portfolio – containing orders and other top-secret documents – lay undisturbed. Any Navy officer who deliberately allowed his confidential portfolio to get out of his hands was risking a general court-martial!

'What really gave me creeps, though,' said one man, 'was this sandwich with one bite out of it and a mug of coffee that had spilled on papers on the desk. The papers were like a blotter and the coffee was still warm!'

That afternoon, the sun burned away the overcast and the sea was as smooth as glass. Moffett Field was reserving judgment on the fate of the *L-8*'s crew until all reports were in from the air-and-sea rescue teams, which were combing the route taken by the blimp from the time it had left its mast on Treasure Island at 6.00

209

a.m. until its ghostly landing in Daly City. Even if the crewmen had drowned at sea, their lifejackets would probably keep their bodies afloat. And, if for some reason the two crewmen had abandoned the blimp on the California coast, sooner or later they would contact the base. They were respected officers, veterans of the airship service.

It was solid Navy logic – but logic did not prevail.

To this day, the crew of the blimp *L-8* has never been heard from. It's as though they had vanished from the face of this planet – as some actually suggest.

*Pan Am pioneered transoceanic transport flying in the 1930s with
their flying boats. By 1945 transatlantic flying by landplanes, four-
engined bombers and transport aircraft, was an everyday thing.
Their navigators proceeded by dead reckoning, supplemented by astro
shots from time to time, and radio compass bearings when within
range of land beacons. Loran came in at the end of the war, and to this
day loran signals have a time delay originally put in to fool the
Germans. There was of course no doppler, no inertial navigation
systems or vortacs. And if you didn't yet carry loran, if the winds you
met were very different from those forecast, and if solid clag stopped
you taking star shots, you could be in trouble. Here is an account of a
C-54 (military DC-4 Skymaster) flight that was in trouble.*

33

FROM: *Song of the Sky*
BY Guy Murchie
Secker & Warburg, London, 1955

An excellent example of the influence of mysterious
and spiritual things on navigation occurred about ten years ago to
a navigator friend of mine named Willie Leveen. Willie was
inadvertently cast in the leading role in a living nightmare over
the British Isles one stormy night near the end of World War II.
His was a true-life adventure which may well take its proper place
in history as a modern navigation classic.

Willie had been my radio instructor in the last months of 1943
and he had served American Airlines as a crack radio-operator
during most of the war. He had learned navigation only in the
early months of 1944. The occasion of his great adventure was his
fifth transoceanic trip as a navigator, under the Air Transport
Command, when he was flying a cargo of 'eighteen American
generals' from the Azores to Prestwick. The Battle of the Bulge
was less than a month away and this heavy helping of high brass
was just returning from a final conference with General Marshall
at the Pentagon.

Willie's weather-folder showed a severe cold front approaching Britain from the west but it was not due to hit Scotland for at
least an hour after the plane's arrival there, so he was not

particularly concerned. The long afternoon dragged on uneventfully until the sun, seen withershins from the aeroplane, swiftly settled into the western ocean. Then suddenly night closed down like an eyelid upon the seeing earth, Ireland was still three hours away and, as the first stars appeared, Willie could see a dense cloud bank far ahead – the rear of the cold front.

Knowing that he had little time left in which to get a celestial position, he quickly picked out several well-dispersed stars and shot himself a good four-star fix. The position showed the aeroplane practically on course and substantiated his 1.05 a.m. ETA on Prestwick.

By the time Willie had worked out the fix it was nearly ten o'clock and the aeroplane had entered the cloud bank. There was now nothing left to navigate with but the radio and dead reckoning. The flight was still going according to plan, however, and neither Willie nor his pilot, Captain Daniel L. Boone, had any apprehension of serious trouble. One of the generals sitting in the passenger's cabin was amusing himself by keeping track of the headings of the aeroplane, aided by his own pocket compass and watch. The aeroplane was a Douglas C-54 and her code name was Great Joy Queen.

After about an hour Willie got a radio bearing from the range station at Valley, in northern Wales. It gave him a line of position that plotted at right angles to his course, an indication of groundspeed. But Willie didn't rate it of much value because radio could not be relied on at two hundred miles out, and besides, the line, if correct, showed that Great Joy Queen must have slowed down to an almost absurd degree.

As the next hour passed, Willie kept expecting better radio reception, but neither he nor his radio-operator could raise a thing. They couldn't even get Valley any more. 'Mighty strange,' thought Willie. 'Could the radio be on the blink?' Not likely, as all three radios on board acted the same. An eerie loneliness came over Willie as he looked out into the black nothingness beyond the windows, and heard only the sound of sizzling fat in his earphones.

He leaned over Captain Boone's shoulder: 'We still can't get a thing on the radio, Dan. All I've got is dead reckoning. Do you think we could climb out of this soup and get a star shot?'

212

'Not a chance, Willie,' said Boone. 'This cumulus stuff goes way up. Better stick to D.R. and keep trying the radio.'

So Willie kept to his original flight plan, using dead reckoning, guessing the wind from the weather folder carried all the way from the Azores. He also kept at the radio. He worked the command set while the radio operator worked the liaison set. Between them they tried the automatic radio compasses too, even the loop. They tried everything. But, as Willie said afterward, 'no dice'.

Landfall had been estimated for about 12.00 p.m., so after midnight Willie assumed Ireland was below. What else could he do? And when the time came, according to flight plan and dead reckoning, he made the turn over the range station at Nutts Corner, but when you don't *know* something in such a case you have to *assume* something. It's at least a hypothesis until proven or disproven – for you can't stop and ponder when you're moving at 180 knots.

As the Queen flew north-eastward toward a hypothetical Prestwick, and still no radio, Willie wondered what to do next. He had long since passed his point of no return, so there was no chance of going back to the Azores. His alternate destination of Valley was still a possibility but without radio would be no easier to find than Prestwick.

Should he try to descend under the clouds and find Prestwick visually? No. With only a vague idea of where he was and no knowledge of whether there was any room between clouds and earth, descending blindly down into mountainous Scotland would be almost as sensible as diving out of an office window in New York City in hopes of landing in a haystack.

Then what about going up? It offered small hope and would use a lot of petrol, but men in dire straits must grasp for anything. Willie again put it to Boone.

'No,' said the captain. 'There must be something you can get on the radio. Radio's our best bet for getting down, Willie. Even a three-star fix can't show us the way down through this soup – but radio can.'

So Willie and his radio-operator twirled the knobs some more. Was there any station at all on the air? Evidently not one . . . No. Nothing. No – yes, there was one. But it was hard to tell whether it

213

was a voice or dots and dashes. Then it was gone. The static sounded like a crackling fire. For millions of miles outward toward the sun the unseen sky was filled with hydrogen ions whipping downward upon the earth, playing hob with magnetic stability around both temperate zones. Scotland was almost in the band of maximum disturbance.

Willie switched on the radio compass again, tuned to Prestwick. The dead needle started breathing, twitched, and moved. Then it reversed itself, wavered and spun around three times . . .

Willie wished he had radar aboard. Radar might just have worked in a time like this. What a help it would have been to get a radar reflection back from the ground, to feel out those craggy Scottish hills a little. But the lesson of Job, 'Speak to the earth, and it shall teach thee', had not yet been learned by this flying boxcar.

Nor was there any loran on this four-engined sky-horse – loran, the new visual radio that since World War II has widely simplified long-range navigation, permitting quick fixes of position by electronic measurements of micro-second intervals between pairs of synchronized stations.

Willie wasted no thought yearning for this fluoroscopic magic that was already enabling other navigators to home in on special loran charts of hyperbolas in many colours – this ballet of the pine needles, of shimmering sky waves, subsea grass and green fire, of storms and snakes and music and lightning standing still. He knew that he couldn't have got a fix on the best loran set in the world under the magnetic pandemonium now enveloping his world. Loran was too new, too delicate, too tricky still – and anyhow he didn't have it.

So, 'What about it, Dan?' asked Willie once more. 'Want to go up? Not a prayer on the radio.'

'Let's try the Irish Sea,' said Captain Boone. 'That's right here somewhere to the west of us. It would be pretty safe to let down there to a couple of thousand feet and maybe we can get on contact and see our way into Prestwick.'

'All right,' said Willie. 'Better fly two-seventy. I don't like this much.'

Boone adjusted the throttle-knobs and started letting down,

turning to a course of 270°. Willie watched anxiously.

Down and down – . . . 8,000 feet . . . 5,000 feet . . . 3,000 feet . . . 2,000 . . . 1,500 . . . Finally Boone levelled off, but still there was no bottom to the clouds.

'You win, Willie,' he said. 'I guess we have no choice now.'

He set his throttles for a long climb. Presently all the men adjusted their oxygen-masks and opened the valves for higher altitude. Willie and the radio-operator kept their earphones on, kept trying everything in the book – but heard only the wail of the unknown void around them, the unearthly howl of outraged electrons flying from the sun. And the altimeter needle moved slowly upwards: 15,000 feet . . . 16,000 . . . 17,000 . . .

After a long half hour Great Joy Queen was getting close to her ceiling, and still in the clouds which seemed to have no end. The needle read 25,200 feet and the big plane was beginning to mush. There was hardly enough air to hold her up, but somehow she managed to claw her way among the molecules of nitrogen still a little higher – 25,400 – and a little higher – 25,500 . . . 25,550 . . .

Willie was beside himself with anxiety, and consciously appealed to what Divine powers there might be in the great unknown vastness above and all around. Could he have a peek at a star? Just a few seconds of a star? Just one little star. Any star would do. Anything would do, please God.

As Dan Boone laboured toward the last inch of ceiling, Willie's grey eyes scanned the dark nothingness out of the astrodome – wistfully, pleadingly, desperately. Was there a light anywhere? A whisper of a star? Now was the crucial time. Now, God.

What was that over there to the east? The frost on the dome? Willie rubbed the frozen breath with his sleeve. And there was still something light up there: the moon!

Ah! Willie thanked God in his heart as he reached for his octant and swung it toward the hazy glow of light. It was the full moon. It was dim and vague but strangely big – as big as a parson's barn – almost too big. The sky was still deep in clouds but now and then Willie could see its roundness clear enough to shoot. He quickly removed his oxygen-mask to clear his face for the eyepiece.

'Hold her steady, Dan,' he called as he balanced the silver bubble and pressed his trigger on the moon. It was the most difficult shot Willie had ever made – and the most fraught with consequence. The angle twisted his neck and he was cold. Besides, he could hardly make out the moon's limb and he had to keep rubbing the frost off the Plexiglass every fifteen seconds, the while dancing on his little stool. He didn't have the traditional electric hair dryer for cooking the frost off the dome. He just rubbed with one hand, desperately. Without oxygen, his breath came in short gasps.

Somehow he managed it, and as Boone started descending again Willie figured and plotted a moon line. But as his fingers drew the line Willie's eyes widened with amazement. The line was mostly off the map. It ran north and south and put the plane somewhere just off the coast of Norway!

'Dan, do a one-eighty turn and let down,' gasped Willie. He half expected Boone to question his wild request, but Boone promptly banked the plane into a complete reversal of direction and the Queen was headed southwest, presumably back across the North Sea toward Britain again. 'We must be in one hell of a west wind,' muttered Willie.

When he was asked later by investigators why he accepted that single implausible moon shot as accurate, Willie replied, 'It was all I had had to go by in more than four hours. What else could I believe?'

Fortunate it was for the war and at least two dozen lives that Willie had that much confidence in himself, and that Boone trusted him too – for fate was figuring close that night, and there were only a couple of hours of fuel left.

As the plane descended steadily toward what Willie presumed was the North Sea, Boone throttled down his engines to save every possible drop of fuel. He put her on maximum range. That is the slow overdrive prop and throttle setting of lean mixture (more air and less petrol) originated by Pan American, with Lindbergh as advisor, for just such an emergency. It means flying just slow enough to squeeze as many miles out of each gallon as you can without mushing.

The plan was to try to get below the clouds and the strong winds while over the sea where it would be reasonably safe to descend that low. To do this Willie had to bet the lives of all on board on his moon shot. The war in France might also feel the consequences. He had to wager everything on coming down to

216

the sea rather than into rugged Norway, Scotland, or the Orkney Isles.

Down, down they went. When the altimeter showed 500 feet, anxiety became intense. If over land, this altitude could easily be disastrous – and Willie could not even judge the accuracy of the altimeters because he had received no barometric correction since the Azores.

At '400 feet' a greyness appeared in the black below. The sea! Willie relaxed a little. Boone levelled off at 200 feet where he could avoid the full force of the evident headwinds of higher levels. Willie gazed anxiously at the water. He thought he could see huge white caps, a gale blowing from the west.

Consulting with Boone, Willie had determined to fly west until the British coast appeared, then fly along the coast in an attempt to recognize some locality and, if possible, find a landing-field. Meanwhile the flight-clerk and engineer were making preparation for possible landing in the sea. Life rafts were dragged forth and Mae West jackets handed to all the generals. It is interesting to think of the comments that must have come from the brass as they were being assigned individually to rafts. Of that, alas, I have no record.

After half an hour Boone suddenly cried, 'Land!' He banked to the right and headed up the coast. All the crew looked eagerly at the dim outline of the shore and Willie tried to match it with some part of his map. It was tantalizing. He could not recognize anything, nor tell whether the coast was England, Scotland, the Shetlands or even something else. Willie felt cold shivers in his bones.

Soon realizing the unlikelihood of identifying the blacked-out coast in time to do any good, Willie decided it would be better to go inland in search of airfields and possible radio contact. So he got Boone to turn west, and they agreed to fly inland for thirty minutes. If they did not discover anything useful in that time they would return to the coast. By then, they figured, the fuel tanks would be about empty. They planned in the end to ditch in the ocean as close to land as possible in hopes of being able to make the shore in their rubber rafts against the gale blowing out to sea.

As they flew west Willie and the radio-operator continued

trying everything in the book on their radios, desperately seeking even the faintest recognizable response. And Willie peered ahead at the same time over Boone's shoulder watching the murky landscape below for a light or a city, a railway line, a highway, a lake – any clue.

At one point Willie suddenly saw a high hill approach dead ahead. It was so close he was sure they would crash. He braced himself frantically as Boone zoomed upward and the 'hill' burst all round them! It was a black cloud – and in four seconds they were out again on the other side. Hard on the nerves, this.

Every now and then Willie would look at the radio compass – just in case it should settle down and come to the point. It was still spinning now and wavering except when passing through clouds, he noticed. Sometimes large cumulus clouds have enough current in them to activate the radio compass and thunderstorms have been known to masquerade as range-stations. So he watched and checked and waited for identification – and listened – and looked some more.

What was that whine in the headphones? Was it Scotland or Norway or Russia? Willie could not decide whether the radio sounded more like bagpipes or Tchaikovsky's Chinese dance. It would have been funny if only it had not been so serious.

When the allotted thirty minutes were nearly gone, the radio-operator suddenly shouted, 'Prestwick! I've got Prestwick!' It was now 4.30 a.m. and this was the first radio contact made in five and a half hours. The spluttering code sang forth as in the Psalm: 'He spake to them in the cloudy pillar'. Willie prayed it would not prove too late.

The radio-operator tapped out a request for position. A couple of minutes passed while Prestwick control and other co-ordinated stations took simultaneous bearings on the aeroplane; then the position was given in exact latitude and longitude.

Willie scribbled it down frantically: '3° 35' W., 53° 20' N.

'Dan,' he shouted, 'do a one-eighty. We're headed for Ireland. We're over the Irish Sea near Liverpool.'

Willie had to think hard. He knew where he was at last, but there was so little fuel left that it seemed out of the question to try to reach Prestwick. Some nearer field would have to be found. But the radio was still scarcely usable and very uncertain.

As the Queen approached land again Willie racked his brains for ideas. He remembered vaguely having heard of an emergency radio system the R.A.F. used for helping disabled bombers find their way home. It was known by the code name of 'Darkee'. It was the emergency Darkee System, but how could Willie find it? What was the frequency?

Willie found himself praying again. 'Dear God, we need You still.' There was not a minute to lose. And to Willie's amazement an answer popped into his head at once: 4,220 kilocycles. 'It came straight from the Lord,' he told me afterward.

Willie's fingers twirled the knobs to 4,220 and held down the microphone button: 'Darkee, Darkee, Darkee—'

He got an answer: 'This is Darkee! Circle. Circle. We are tracking you . . . Now we have you. Fly one-twenty degrees. We will give you further instructions. Altimeter setting is 29·31. Highest obstruction four hundred feet.'

Willie leaned over Boone's shoulder as Boone flew the course of 120°. He corrected the altimeters to 29·31 for existing barometric pressure, and Boone kept the Queen at 600 feet. It was so dark that scarcely anything of the landscape below could be seen and often it was obscured by fog or low clouds. Time went by . . . fifteen minutes . . . twenty minutes . . .

Just when Willie was beginning to expect splutters from the engines as the fuel-tanks went dry, Darkee said: 'Make a three-sixty turn. You are over the field. Let down to five hundred feet.'

Boone did as he was told, but could see nothing of the ground. 'Darkee, we are still in solid clouds,' he reported.

'Then go down to four hundred', said Darkee.

When even that failed to reveal the ground, Darkee urged, 'Three hundred feet but very carefully.'

Again Boone crept downward, feeling his way with eyes now on the altimeter, now on the blackness beyond the windshield. When the needle read 300, the clouds remained as impenetrable as ever.

'Still can't see you, Darkee.'

'I can hear you plainly,' said Darkee. 'You are south of the tower now, about two miles. Fly thirty degrees. That will bring you over the tower.'

Boone banked quickly until his compass showed 30°. In less

than a minute Darkee said, 'Now you are exactly above me. Circle to the right and let down to two hundred feet.'

Boone nosed downward again and at 200 feet saw what seemed to be an opening in the murk. Venturing to 150 feet he could dimly make out the ground at times, but no sign of an airport.

'Where is the runway, Darkee? Will you shoot off a flare for us?'

A beautiful green flame rocketed into the sky from almost directly below.

'We are right over you, Darkee, but can't see any runway.'

'The runway is below you now. Circle and land! You will see it. Circle and land!'

As Boone circled desperately once more, Willie noticed that the fuel gauges read zero. Still no runway in sight. Long afterward Boone was to write to me: 'When I think about it I get a sick feeling in the pit of my stomach.'

'We are going to land anyway,' cried Boone. 'Give us all the lights you've got, Darkee. We are out of gas. We have no choice.'

Just then there was a sputter from number four engine and it quickly died. Boone circled to the left – descending – apprehensively searching. Suddenly two lines of lights appeared below. The runway! By some strange quirk of mind Darkee had forgotten in his excitement to turn on the lights until now. Skilfully swinging around to line up with the runway, Daniel Boone put her gear down and brought the Queen in on three engines, easing her steeply into the little field, an R.A.F. fighter-base. The wheels touched, bounced, She was rolling fast and the runway was short. The brakes squealed and smoked and Boone pulled his emergency bottle, a hydraulic device that locks the brakes – something to be used only under desperate circumstances.

When the big plane finally screeched to rest at the very end of the runway, Boone swung her around to taxi to the ramp. It was only then that Willie noticed that number three engine was also dead. And by the ramp the other two engines had started to sputter. The tanks were dry. They had landed in the little town of Downham Market, eighty miles north of London.

When I last saw Willie a few months ago, he was just out of the hospital after a nearly fatal accident in which his car was hit by a big grocery truck. He had broken a leg and an arm, several ribs, fractured his skull badly and, as in the adventure over Britain, had escaped only with the skin of his soul.

As we walked across Union Square in New York City to lunch I noticed Willie would not venture a foot from the kerb until the lights were indubitably in our favour. 'I'm not taking any chances,' said Willie reverently. 'God has always pulled me through the pinches, and I'm not gonna put undue strain on our relations.'

Jets arrived too late to have much effect in World War Two, but there have been some big jet battles since, Korea and Vietnam and during the Arab-Israeli wars. Here is an account of a fast-moving flight between Russian MiGs and American Sabrejets over North Korea. This was a true religious war in the old medieval traditions: our belief in capitalism and the free society against yours in Marx and social equality. It was an old man's war: the top-scoring US fighter pilots in Korea were mostly around forty years of age, veterans of World War Two.

34

MiG Alley
By Colonel Harrison R. Thyng
FROM: Air Power, the Decisive Force in Korea
EDITED BY James T. Stewart
Van Nostrand, London, 1957

Like olden knights the F-86 pilots ride up over North Korea to the Yalu River, the sun glinting off silver aircraft, contrails streaming behind, as they challenge the numerically superior enemy to come on up and fight. With eyes scanning the horizon to prevent any surprise, they watch avidly while MiG pilots leisurely mount their cockpits, taxi out onto their runways for a formation take-off.

'Thirty-six lining up at Antung,' Black Leader calls.

'Hell, only twenty-four taking off over here at Tatungkou,' complains Blue Leader.

'Well, it will be at least three for everybody. I count fifty at Takushan,' calls White Leader.

'I see dust at Fen Cheng, so they are gathering up there,' yells Yellow Leader.

Once again the Commie leaders have taken up our challenge, and now we may expect the usual numerical odds as the MiGs gain altitude and form up preparatory to crossing the Yalu.

Breaking up into small flights, we stagger our altitude. We have checked our guns and sights by firing a few warm-up rounds as we crossed the bomb line. Oxygen masks are checked and pulled as tight as possible over our faces. We know we may exceed eight

'Gs' in the coming fight, and that is painful with a loose mask. We are cruising at a very high Mach. Every eye is strained to catch the first movement of an enemy attempt to cross the Yalu from their Manchurian sanctuary into the graveyard of several hundred MiGs known as 'MiG Alley'. Several minutes pass. We know the MiG pilots will become bolder as our fuel time limit over the 'Alley' grows shorter.

Now we see flashes in the distance as the sun reflects off the beautiful MiG aircraft. The radio crackled, 'Many, many coming across at Suiho above forty-five thousand feet.' Our flights start converging toward that area, low flights climbing, yet keeping a very high Mach. Contrails are now showing over the Antung area, so another section is preparing to cross at Sinuiju, a favourite spot.

We know the enemy sections are now being vectored by GCI, and the advantage is theirs. Travelling at terrifically high speed and altitude, attackers can readily achieve surprise. The area bound by the horizon at this altitude is so vast that it is practically impossible to keep it fully covered with the human eye.

Our flights are well spread out, ships line abreast, and each pilot keeps his head swivelling 360 degrees. Suddenly MiGs appear directly in front of us at our level. At rates of closure of possibly 1,200 miles an hour we pass through each other's formations.

Accurate radar ranging firing is difficult under these conditions, but you fire a burst at the nearest enemy anyway. Immediately the MiG's zoom for altitude, and you break at maximum 'G' around toward them. Unless the MiG wants to fight and also turned as he climbed, he will be lost from sight in the distance before the turn is completed. But if he shows an inclination to scrap, you immediately trade head-on passes again. You 'sucker' the MiG into a position where the outstanding advantage of your aircraft will give you the chance to outmanoeuvre him.

For you combat has become an individual 'dogfight'. Flight integrity has been lost, but your wing man is still with you, widely separated but close enough for you to know that you are covered. Suddenly you go into a steep turn. Your Mach drops off. The MiG turns with you, and you let him gradually creep up and out-turn you. At the critical moment you reverse your turn. The hydraulic controls work beautifully. The MiG cannot turn as readily as you and is slung out to the side. When you pop your speed brakes, the MiG flashes by you. Quickly closing the brakes, you slide on to his tail and hammer him with your '50's'. Pieces fly off the MiG, but he won't burn or explode at that high altitude. He twists and turns and attempts to dive away, but you will not be denied. Your '50's' have hit him in the engine and slowed him up enough so that he cannot get away from you. His canopy suddenly blows and the pilot catapults out, barely missing your airplane. Now your wing man is whooping it up over the radio, and you flash for home very low on fuel.

At this point your engine is running very rough. Parts of the ripped MiG have been sucked into your engine scoop, and the possibility of its flaming out is very likely. Desperately climbing for altitude you finally reach forty thousand feet. With home base now but eighty miles away, you can lean back and sigh with relief for you know you can glide your ship back and land, gear down, even if your engine quits right now. You hear over the radio, 'Flights reforming and returning – the last MiG chased back across the Yalu.' Everyone is checking in, and a few scores are being discussed.

The good news of no losses, the tension which gripped you before the battle, the wild fight, and the 'G' forces are now being felt. A tired yet elated feeling is overcoming you, although the day's work is not finished. Your engine finally flames out, but you have maintained forty thousand feet and are now but twenty miles from home. The usual radio calls are given, and the pattern set up for a dead-stick landing. The tower calmly tells you that you are number three dead-stick over the field, but everything is ready for your entry. Planes in front of you continue to land in routine and uninterrupted precision, as everyone is low on fuel. Fortunately this time there are no battle damages to be crash landed. Your altitude is decreasing, and gear is lowered. Hydraulic controls are still working beautifully on the pressure maintained by your wind-milling engine. You pick your place in the pattern, land, coast to a stop, and within seconds are tugged up the taxi strip to your revetment for a quick engine change.

Debriefing begins at once, and the excitement is terrific as the score for the mission mounts to four MiGs, confirmed, one

d. A quick tally discloses that we had
 ast three to one, but once again the enemy
 cked up.

 e type most enjoyed by the fighter pilot. It is a
 weep, with no worries about escort or providing
 er-bombers. The mission had been well planned
 cuted. Best of all, the MiGs had come forth for battle.
 ate flights had probably again confused the enemy
 pe readers, and, to an extent, nullified that tremendous
 advantage which radar plotting and vectoring gives a
 er on first sighting the enemy. We had put the maximum
 umber of aircraft into the target area at the most opportune time,
and we had sufficient fuel to fool the enemy. Our patrolling flights
at strategic locations had intercepted split-off MiGs returning
towards their sanctuary in at least two instances. One downed
MiG had crashed in the middle of Sinuiju, and another, after
being shot-up, had outrun our boys to the Yalu, where they had to
break off pursuit. But they had the satisfaction of seeing the
smoking MiG blow up in his own traffic pattern. Both instances
undoubtedly did not aid the morale of the Reds.

Richard Bach is the dreamer who wrote that
Livingston Seagull, *an allegory about the*
determination, in the form of an account of a s
own frontiers of flight. The book was turned
publishers before finding a home, where it earned ev
deal of money for its author. Before than, Bach was an a
of Flying *magazine, as I was at that time; I was bowle*
skill at writing. Earlier, he had been a sometime pilot in

Of all his books, I best like not Seagull *but his first, Str*
the Ground, *a narrative of a flight he had made one thundery*
as a pilot of an F-84F Thunderstreak jet, carrying a pouch of se
papers from one US air base in England to another in France. It was
very nearly the end of Bach and Thunderstreak, for he lost the
aeroplane twice, once inside a thundercloud. If you do get a jet
coming downhill fast and you are on instruments, perhaps with the
gyros tumbling, you have only a short moment in which to recover
before it is too late. 'Jet upsets' like this have brought down even
airliners on occasion.

Here is his description of those minutes so near to death.

35

FROM: *Stranger to the Ground*
BY   Richard Bach
Cassell, London, 1963

How slowly it is, though, that we learn of the nature of dying. We form our preconceptions, we make our little fancies of what it is to pass beyond the material, we imagine what it feels like to face death. Every once in a while we actually do face it.

It is a dark night, and I am flying right wing on my flight leader. I wish for a moon, but there is none. Beneath us by some six miles lie cities beginning to sink under a gauzy coverlet of mist. Ahead the mist turns to low fog, and the bright stars dim a fraction in a sheet of high haze. I fly intently on the wing of my leader, who is a pattern of three white lights and one of green. The lights are too bright in the dark night, and surround themselves with brilliant flares of halo that make them painful to watch. I press the microphone button on the throttle. 'Go dim on your nav lights, will you, Red Leader?'

'Sure thing.'

In a moment the lights are dim, mere smudges of glowing filament that seek more to blend his airplane with the stars than to set it apart from them. His airplane is one of the several whose *dim* is just too dim to fly by. I would rather close my eyes against the glare than fly on a shifting dim constellation moving among the brighter constellations of stars. 'Set'em back to bright, please. Sorry.'

'Roj.'

It is not really enjoyable to fly like this, for I must always relate that little constellation to the outline of an airplane that I know is there, and fly my own airplane in relation to the mental outline. One light shines on the steel length of a drop tank, and the presence of the drop tank makes it easier to visualize the airplane that I assume is near me in the darkness. If there is one type of flying more difficult than dark-night formation, it is dark-night formation in weather, and the haze thickens at our altitude. I would much rather be on the ground. I would much rather be sitting in a comfortable chair with a pleasant evening sifting by me. But the fact remains that I am sitting in a yellow-handled ejection seat and that before I can feel the comfort of any evening again I must first successfully complete this flight through the night and through whatever weather and difficulties lie ahead. I am not worried, for I have flown many flights in many airplanes, and have not yet damaged an airplane or my desire to fly them.

France Control calls, asking that we change to frequency 355·8. France Control has just introduced me to the face of death. I slide my airplane away from leader's just a little, and divert my attention to turning four separate knobs that will let me listen, on a new frequency, to what they have to say. It takes a moment in the red light to turn the knobs. I look up to see the bright lights of Lead beginning to dim in the haze. I will lose him. Forward on the throttle, catch up with him before he disappears in the mist. Hurry.

Very suddenly in the deceptive mist I am closing too quickly on his wing and his lights are very very bright. Look out, you'll run right into him! He is so helpless as he flies on instruments. He couldn't dodge now if he knew that I would hit him. I slam the throttle back to *idle*, jerk the nose of my airplane up, and roll so that I am upside down, watching the lights of his airplane through the top of my canopy.

Then, very quickly, he is gone. I see my flashlight where it has fallen to the plexiglass over my head, silhouetted by the diffused yellow glow in the low cloud that is a city preparing to sleep on the ground. What an unusual place for a flashlight. I begin the roll to recover to level flight, but I move the stick too quickly, at what has become far too low an airspeed. I am stunned. My airplane is spinning. It snaps around once and the glow is all about me. I look for references, for ground or stars; but there is only the faceless glow. The stick shakes convulsively in my hand and the airplane snaps around again. I do not know whether the airplane is in an erect spin or an inverted spin, I know only that one must never spin a swept-wing aircraft. Not even in broad light and clear day. Instruments. Attitude indicator shows that the spin has stopped, by itself or by my monstrous efforts on the stick and rudder. It shows that the airplane is wings-level inverted; the two little bars of the artificial horizon that always point to the ground are pointing now to the canopy overhead.

I must bail out. I must not stay in an uncontrolled airplane below 10,000 feet. The altimeter is an unwinding blur. I must raise the right armrest, squeeze the trigger, before it is too late.

There is a city beneath me. I promised myself that I would never leave an airplane over a city.

Give it one more chance to recover on instruments, I haven't given the airplane a chance to fly itselt out.

The ground must be very close.

There is a strange low roaring in my ears.

Fly the attitude indicator.

Twist the wings level.

Speed brakes out.

I must be very close to the ground, and the ground is not the friend of airplanes that dive into it.

Pull out.

Roaring in my ears. Glow in the cloud around me.

St. Elmo's fire on the windscreen, blue and dancing. The last time I saw St. Elmo's fire was over Albuquerque, last year with Bo Beaven.

Pull out.

Well, I am waiting, death. The ground is very close, for the glow is bright and the roaring is loud. It will come quickly. Will I hear it or will everything just go black? I hold the stick back as hard as I dare – harder would stall the airplane, spin it again.

So this is what dying is like. You find yourself in a situation that has suddenly gone out of control, and you die. And there will be a pile of wreckage and someone will wonder why the pilot didn't eject from his airplane. One must never stay with an uncontrolled airplane below 10,000 feet.

Why do you wait, death? I know I am certain I am convinced that I will hit the ground in a few thousandths of a second. I am tense for the impact. I am not really ready to die, but now that is just too bad. I am shocked and surprised and interested in meeting death. The waiting for the crash is unbearable.

And then I am suddenly alive again.

The airplane is climbing.

I am alive.

The altimeter sweeeps through 6,000 feet in a swift rush of a climb. Speed brakes *in*. Full forward with the throttle. I am climbing. Wings level, airspeed a safe 350 knots, the glow is fading below. The accelerometer shows that I pulled seven and a half G's in my recovery from the dive. I didn't feel one of them, even though my G-suit was not plugged into its source of pressured air.

'Red lead, this is Two here; had a little difficulty, climbing back through 10,000 feet . . .'

'Ten thousand feet?'

'Roger, I'll be up with you in a minute, we can rejoin over Toul TACAN.'

Odd. And I was so sure that I would be dead.

The flashes in the dark clouds north of Phalsbourg are more frequent and flicker now from behind my airplane as well as in front of it. They are good indicators of thunderstorm cells, and they do not exactly fit my definition of 'scattered.' Directly ahead, on course, are three quick bright flashes in a row. Correct

30 degrees left. Alone. Time for twisted thoughts in the back of the mind. 'You have to be crazy or just plain stupid to fly into a thunderstorm in an eighty-four F.' The words are my words, agreed and illustrated by other pilots who had circumstance force them to fly this airplane through an active storm cell.

The airplane, they say, goes almost completely out of control, and despite the soothing words of the flight handbook, the pilot is relying only on his airplane's inertia to hurl it through and into smooth air beyond the storm.

But still I have no intention of penetrating one of the flickering monsters ahead. And I see that my words were wrong. I face the storms on my course now through a chain of logic that any pilot would have followed. The report called them 'scattered,' not numerous or continuous. I fly on. There are at least four separate radar-equipped facilities below me capable of calling vectors through the worst cells. I fly on. A single-engine pilot does not predicate his action on what-shall-I-do-if-the-radio-goes-out. The risk of the mission is worth the result of delivering the heavy canvas sack in the gun bay.

Now, neither crazy nor stupid, I am at the last link of the chain: I dodge the storms by the swerving radiocompass needle and the flashes of lightning that I see from the cockpit. The TACAN is not in the least disturbed by my uneasy state of mind. The only thing that matters in the world of its transistorized brain is that we are 061 miles from Phalsbourg, slightly to the left of course. The radio-compass has gone wild, pointing left and right and ahead and behind. Its panic is disconcerting among the level-headed coolness of the other instruments, and my right glove moves its function switch to *off*. Gratefully accepting the sedative, the needle slows, and stops.

Flash to the left, alter course 10 degrees right. Flash behind the right wing, forget about it. Flash-flash directly brilliantly ahead and the instrument panel goes featureless and white. There is no dodging this one. Scattered.

The storm, in quick sudden hard cold fury, grips my airplane in its jaws and shakes it as a furious terrier shakes a rat. Right glove is tight on the stick. Instrument panel, shock-mounted, slams into blur. The tin horizon whips from an instant 30-degree left bank to an instant 60-degree right bank. That is not possible. A storm is only air.

230

Left glove, throttle full forward. My airplane, in slow motion, yaws dully to the left. Right rudder, hard. Like a crash landing on a deep-rutted rock trail. Yaw to the right. My airplane has been drugged, she will not respond. Vicious left rudder.

The power, where is the power? Left glove back, forward again, as far as it will go, as hard as it will go. A shimmering blurred line where the tachometer needle should be. Less than 90 per cent rpm at full throttle.

I hear the airplane shaking. I cannot hear the engine. Stick and rudders are useless moving pieces of metal. I cannot control my airplane. But throttle, I need the throttle. What is wrong?

Ice. The intake guide vanes are icing, and the engine is not getting air. I see intake clogged in grey ice. Flash and FLASH the bolt is a brilliant snake of incandescent noon-white sun in the dark. I cannot see. Everything has gone red and I cannot even see the blurred panel. I feel the stick I feel the throttle I cannot see. I have suddenly a ship in the sky, and the storm is breaking it. So quickly. This cannot last. Thunderstorms cannot hurt fighters. I am on my way to Chaumont. Important mission.

Slowly, through the bone-jarring shake of the storm, I can see again. The windscreen is caked with grey ice and bright blue fire. I have never seen the fire so brightly blue. My wings are white. I am heavy with ice and I am falling and the worst part of a thunderstorm is at the lowest altitudes. I cannot take much more of this pounding. White wings, covered in shroud. Right glove grips the stick, for that is what has kept my airplane in the sky for six years. But tonight the airplane is very slow and does not respond, as if she were suddenly very tired and did not care to live. As if her engine had been shut down.

The storm is a wild horse of the desert that has suddenly discovered a monster on its back. It is in a frenzy to rid itself of me, and it strikes with shocks so fast they cannot be seen. I learn a new fact. The ejection seat is not always an escape. Bailout into the storm will be just as fatal as the meeting of earth and airplane, for in the churning air my parachute would be a tangled nylon rag. My airplane and I have been together for a long time, we will stay together now. The decision bolts the ejection seat to the cockpit floor, the *Thunderstreak* and I smash down through the jagged sky as a single dying soul. My arm is heavy on the stick, and tired. It

231

will be good to rest. There is a roaring in my ears, and I feel the hard ground widening about me, falling up to me.

So this is the way it will end. With a violent shuddering of airplane and unreadable instrument panel; with a smothered engine and heavy white wings. Again the feeling: I am not really ready to end the game. I have told myself that this day would come to meet me, as inevitably as the ground which rushes to meet me now, and yet I think, quickly, of a future lost. It cannot be helped. I am falling through a hard splintering storm with a control stick that is not a control stick. I am a chip in a hurricane a raindrop in a typhoon about to become one with the sea a mass of pieces-to-be a concern of air traffic controllers and air police and gendarmerie and coroners and accident investigators and statisticians and newspaper reporters and a board of officers and a theatre commander and a wing commander and a squadron commander and a little circle of friends. I am a knight smashed from his square and thrown to the side of the chessboard.

Tomorrow morning there will be no storm and the sun will be shining on the quiet bits of metal that used to be Air Force Jet Two Niner Four Zero Five.

But at this instant there is a great heavy steel-bladed storm that is battering and crushing me down, out of the sky, and the thing that follows this instant is another just like it.

Altimeter is a blur, airspeed is a blur, vertical speed is a blur, attitude indicator is a quick-rocking blurred luminous line that does not respond to my orders. Any second now, as before, I am tense and waiting. There will be an impact, and blackness and quiet. Far in the back of my mind, behind the calm fear, is curiosity and a patient waiting. And a pride. I am a pilot. I would be a pilot again.

The terrier flings the rat free.

The air is instantly smooth, and soft as layered smoke. Altimeter three thousand feet airspeed one-ninety knots vertical speed four thousand feet per minute down attitude indicator steep right bank heading indicator one seven zero degrees tachometer eighty-three per cent rpm at full throttle. Level the white wings. Air is warm. Thudthudthud from the engine as ice tears from guide vanes and splinters into compressor blades. Wide slabs of ice rip from the wings. Half the windscreen is

232

suddenly clear. Faint blue fire on the glass. Power is taking hold: 90 per cent on the tachometer . . . thud . . . 91 per cent . . . thudthud . . . 96 per cent. Airspeed coming up through 240 knots, left turn, climb. Five hundred feet per minute, 700 feet per minute altimeter showing 3,000 feet and climbing I am 50 degrees off course and I don't care attitude indicator showing steady left climbing turn I'm alive the oil pressure is good utility and power hydraulic pressure are good I don't believe it voltmeter and loadmeter showing normal control stick is smooth and steady how strange it is to be alive windscreen is clear thud 99 per cent rpm tailpipe temperature is in the green. Flash-FLASH look out to the left look out! Hard turn right I'll never make it through another storm tonight forget the flight plan go north of Phalsbourg 15,000 feet 320 knots flash to the left and behind, faint.

And strangely, the words of an old pilot's song: ". . . for I, am too young, to die . . .' It is a good feeling, this being alive. Something I haven't appreciated. I have learned again.

Rpm is up to 100 per cent. I am climbing, and 20,000 feet is below flash 21,000 feet is below. Blue fire washes across the windscreen as if it did not know that a windscreen is just a collection of broken bits of glass.

What a ridiculous thought. A windscreen is a windscreen, a solid piece of six-ply plate glass, for keeping out the wind and the rain and the ice and a place to look through and a place to shine the gunsight. I will be looking through windscreens for a long time to come.

Why didn't I bail out? Because the seat was bolted to the cockpit floor. No. Because I decided not to bail out into the storm. I should have bailed out. I definitely should have left the airplane. Better to take my chances with a rough descent in a torn chute than certain death in a crash. I should have dropped the external tanks, at least. Would have made the airplane lighter and easier to control. Now, at 32,000 feet, I think of dropping the tanks. Quick thinking.

Flash.

I flew out of the storm, and that is what I wanted to do. I am glad now that I did not drop the tanks; there would have been reports to write and reasons to give. When I walk away from my airplane tonight I will have only one comment to make on the Form

One: UHF transmitter and receiver failed during flight. I will be the only person to know that the United States Air Force in Europe came within a few seconds of losing an airplane.

Flashflash. Ahead.

I have had enough storm-flying for one night. Throttle to 100 per cent and climb. I will fly over the weather for the rest of the way home; there will be one cog slipping tonight in the European Air Traffic Control System, above the weather near Phalsbourg. The cog has earned it.

How can they ever get anyone to sit in those centre seats in the Jumbos, so uselessly far from the small windows that you can see nothing outside the claustrophobic cabin? 'Nobody really wants to look out any more,' say the airline marketing men, shuffling their in-depth mass-interview attitude surveys. But surely many of us still count a jet journey as magic if we can sit by a window, and mere time-out-of-life when we can't? So for all who like to gaze out that plastic porthole and daydream away the hours (one hour equals 600 miles), here is this marvellous description of a proving flight made in one of the first and earliest jet airliners, twenty-five years ago.

36

FROM: *No Echo in the Sky*
BY Harald Penrose
Cassell, London, 1958

The year was 1952, but that meant nothing up there except a cipher with which to use the star tables. In the faintly throbbing silence of the flight deck each sat wrapped in thought, sub-consciously dealing with the routine of flying a big four-engined turbine-jet aeroplane at stratospheric heights, half across the world and back. Through the courtesy of De Havilland's I had earlier been allowed a spell at the controls of the Comet, and I had eagerly accepted the invitation to fly on the BOAC familiarization flight to Singapore and back. Enclosed by matt black walls which were patterned by a hundred dark dials, we peered outward past the windscreen, as though from a tunnel's mouth, into the sunlight beyond. The vast dome of unimaginable heights was brushed with black and set upon a wall of silver-frosted blue. Its base formed the remote horizon encircling the microcosm earth lying frail as a fallen leaf upon the oceanic surface of the world.

In spite of the epoch-making swiftness of our flight, height retained the old yet ever new illusion that the aeroplane was poised unmoving in high space, whilst hour by hour the sun's white blaze crossed its track, brushing cloud and sea and land with endless combinations of light and shade. Confirmation that even in the frozen stillness a scale of time should measure distance, came from the speeding minute hand on the dashboard, though ten minutes or an hour of flight could equally well embrace the whole eyespan of space between wilderness and man's salvation.

*   *   *

Through the gauzy curtains of space we stared down at the great areas of jagged mountain tops, and rolling forests; marshy deltas and jungle river courses; pink-tinted deserts of barren sand. But whatever the desolation, fingers of cultivation plucked insecurely at the fringes of the primitive and untamed. The vestiges of earth that man had laboriously smoothed in the thin crust covering the primeval rock were pitifully clear to see: startling evidence that of all the world's land no more than a tenth was a man's heritage wherein to feed 2,000,000,000 inconsequently begetting and multiplying people who dreamed the earth was made for them.

Down below, misted and beautiful, was a world of fantasy, an idiot's dream, a cataclysmic upheaval frozen in its agony, bathed with radiance and painted with delight. Slowly as a cloud travelling across the path of the sun, it passed from sight, and instead there was only the emptiness of sea. Between endless water and unending space we stayed suspended, inanimate, content to wait in the warm, radiant whiteness of the sun, as though at last the deepest secret had been revealed and there was nothing more to seek.

The faint vibration of the whirling turbines buried in the wings flooded like a thin stream through the metallic fibres of the structure. The pulse of its own life beat with ours as the aeroplane persistently aspired to horizons hidden beyond the rim of the world. Presently a purple shadow on the ocean became another land, its roots hidden in the depths of the great waters that are the essence of this world. Soon there was no more sea, and we flew, unimagined and unseen through the high space of yet another continent.

How could I guess that the great stillness was illusion? How could I know that the timeless plain, the ageless desert, the smooth ice-cap, the snow-topped mountain, the unstirring chasm, or the slumbrous island floating on an azure sea, had none of them known quietness? Yet from this great and revealing height, there were many signs, if they could be interpreted, to show that the bones of the planet were not petrified but living – swelling, contracting, shuddering, moving with sluggish pulse. Slowly, infinitely slowly, new mountains were being raised by this dynamic energy, and old ones were being eroded by the rough hand of time, Fissure and faults were forming under the strain of unbearable internal stresses. New lands were appearing, and long-established shores dwindling. Sun, rain, and frost were eternally smoothing the ruckling, heaving skin of sand, chalk, clay, and decay that covered sedimentary rocks long disrupted from their fundamental creation of granite and basalt which encircled the nickel core of our planet. It was a surface world of inexorable and ceaseless change – an earth still tortured, still tormented, built on an unending pattern of destruction and rebirth.

Hour after hour, riding the thin, unbreathable air in the deep suspense of space, we stared from the artificial life of the pressurized cabin at the gradual unfolding of thousand upon thousand miles of natural existence. Only at rare intervals did the high turbulence of great air currents sweeping the skies of the world interrupt the smooth steadiness of the flight. Otherwise the absence of the propellors added to the illusion of timeless trance in which the unrolling panorama of a world was displayed in all its dwarfed nakedness.

Through the armoured side-windows I could see the pointed tips of the tapered wings far astern of the long nose in which we sat. The air intakes of the four jet turbines were hidden by the streamlining of the cabin wall, so that there was no sign of motive power, and the aeroplane seemed to be floating divinely on the high winds of heaven.

*   *   *

On and on we flew, in sunlight and in cloud, by day and by night. Every few hours we descended swiftly to earth, and resumed mortality, half suffocating in the tropic heat, while the great fuel tanks were quickly filled. Then once more into the frozen heights of the upper air, that had to be compressed from thin iciness to burning density, cooled by refrigeration and made humid with moisture to suit our frail lungs.

The drop of the sun below the edge of the land brought night unheralded; and in the tropic darkness the stars by which we checked our flight were pin-point, no longer beautiful with friendly loveliness, nor intimate like moon and earth, but coldly geometric.

Time waited – until dawn caught our flint, a quick flame flickering in the dusky curtain of the east, and draining the sky to cold, bleached blue. Scattered clouds glowed, grew firm and white, and with a single breath the universe threw off the last veil of darkness. Across the edge of space, the horizon cleared and the rim of the sun flooded the earth with daylight.

*   *   *

Far away on earth man might feel the wind and rain, and cower for shelter from the darkening storm that hid the landscape. Viewed from the stratosphere that same storm might be seen shadowing a hundred miles of land beneath white-quilted cloud that could be covered with two fingers. Beyond lay pools of sunshine within reach of a mere fifteen minutes of flying time.

Hours became days whilst we watched new land and sea and sky. A silver insect trapped beneath the domed ceiling of high heaven, our aeroplane drew its evanescent trail swiftly around the earth's circumference. To us it was the very matrix of the universe, the living womb cradling our lives, enclosing crew and passengers protectively with warmth and infinite assurance. Its cabin, trembling faintly, was the only reality in which we could believe. A score of races of mankind must have heard the faint singing of the unseen passage of our stratospheric flight. To them, what could it matter, that a jet airliner heralded another epoch in the history of the earth? The sun and rain, the plots they tilled, their hunger, and their love remained unchanged.

Yet the voice of man continually reached up to us with speech and signal, for we rode the path of wireless waves guiding the long trajectory of flight to the hidden target of each further destination. In a long line forming the image of the earth-track of our skyway, station after station listened for the call sign; heard, reported our bearing; and passed us on. Beacon and track guide, station fix and radar cover, sent vibrations trembling through the air, to be transformed through electronic valves into speech or staccato note or image that linked us to the care of men on earth. But it was a frail thread. The clouds had only to raise their electrically charged heads to distort with static and perjure the signals on which we placed our trust. Even the softness of night could ruffle the rhythm of radio waves into a false echo of their true intent. It was then that the message of the stars speeded our course with certainty as though they were the lighted pinpoints of a map. But when presently from star-fixed space the long descent to earth must be made through the dense opaqueness of many layers of cloud, we flew with tense alertness until radio aids once more reached us without distortion.

Shrouded in blinding mists, the illusory senses of the body relied for balance and direction upon the readings of a multitude of dials. No aerodrome could be located by act of faith in compass and calculation where an error of ten miles was high precision. In clear sunlight, or even unclouded darkness, landmarks by day and lights by night added visual direction to our destination. But of all our journey endings, a full half were made in mist and rain, or with cloud so low above the earth that the restricted space of air between seemed suffocating under the hugeness of our wings. It was then that the guiding signals of radar and ground control were a matter of life and death.

In the tenth hour of the last day's flight the radio led to the final long let-down from high above a cloud-covered France. Within ten minutes a ripple of water gleamed through the mist-blue distance – and suddenly the endearing smallness of southern England grew out of the horizon as though fanned to flame from the spark of faith in our minds. Only a strip of sea and a diminishing gulf of air remained between us. The shores of France slid past the cabin windows and were gone with a seeming whisper that it was not the empty deserts, the lifeless mountains, nor the desolate seas which separated man from man, but the wilderness in human minds and hearts.

Expectantly we descended towards our homes and loves. England was suffused with the level light of evening. Memory of the world's barrenness in ocean wastes and tortured nakedness of land vanished as though we had never speculated on its cruelty. The echo of primitive fear of the earth had died. The impact of vastness and solitude was forgotten. The awareness of the power throbbing through the universe faded. In their train came a sense of relaxation and infinite peace under the spell of England's charm: her gracious sweep of downs; her quiet vales of trim meadows and neat woods; her inimitable quality of greatness and power.

Soon we would land, the long runway hissing under the wheels, the machine settling its weight more and more firmly on the ground until at last it was wholly sustained. I would step again upon the good earth, look about me, breathe the evening air. The rich reality of all the sights and sounds of mankind I had forsaken for the splendid solitude of flight, would return. Presently there would be lamp-light – the candle-flame of life and love, the murmur of voices; the shelter and closeness when darkness would be a velvet curtain of privacy; and the stars sparkling jewels devoid of all significance except their beauty.

Anybody can be brave, but it takes a special type of aviator to be brave at night. The war in Vietnam aroused passions all around the world that will not subside for many years yet; but no one would surely deny the cold courage of the North Vietnamese skulking in the jungles, or of the Americans and South Vietnamese fighting them. Held down by the most massive air power ever organized, the Reds took to moving themselves and their supplies only at night; and the Americans had to go out and try and find their trucks on the Ho Chi Minh trail, also at night.

Here Major Mark E. Berent, serving with the 497th Tactical Fighter Squadron based in Thailand, the only F-4 squadron all-night-flying, vividly describes such a mission. His account first ran in the USAF's Air Force/Space Digest *in 1971.*

37

Night Mission on the Ho Chi Minh Trail

BY   Major Mark E. Berent
USAF *Air Force/Space Digest*, 1971

It's cool this evening, thank God. The night is beautiful, moody, an easy rain falling. Thunder rumbles comfortably in the distance. Just the right texture to erase the oppressive heat memories of a few hours ago. Strange how the Thai monsoon heat sucks the energy from your mind and body by day, only to restore it by the cool night rain.

I am pleased by the tranquil sights and sounds outside the BOQ room door. Distant ramp lights, glare softened by the rain, glisten the leaves and flowers. The straightdown, light rain splashes gently, nicely on the walkways, on the roads, the roofs. Inside the room I put some slow California swing on the recorder (*You gotta go where you wanta go . . .*) and warm some soup on the hot-plate. Warm music, warm smell . . . I am in a different world. (*Do what you wanta, wanta do. . . .*) I've left the door open – I like the sound of the rain out there.

A few hours later, slightly after midnight, I am sitting in the cockpit of my airplane. It is a jet fighter, a Phantom, and it's a good airplane. We don't actually get into the thing – we put it on. I am attached to my craft by two hoses, three wires, lap belt, shoulder harness and two calf garters to keep my legs from flailing about in a highspeed bailout. The gear I wear – gun, G-suit, survival vest, parachute harness – is bulky, uncomfortable, and means life or death.

I start the engines, check the myriad systems – electronic, radar, engine, fire control, navigation – all systems; receive certain information from the control tower, and am ready to taxi. With hand signals we are cleared out of the revetment and down the ramp to the arming area.

I have closed the canopy to keep the rain out, and switch the heavy windscreen blower on and off to hold visibility. I can only keep its hot air on for seconds at a time while on the ground, to prevent cracking the heavy screen. The arming crew, wearing bright colours to indicate their duties, swarm under the plane: electrical continuity – checked; weapons – armed; pins – pulled. Last all-round look-see by the chief – a salute, a thumbs-up, we are cleared. God, the rapport between pilot and ground crew – their last sign, thumbs-up – they are with me. You see them quivering, straining bodies posed forward as they watch *their* airplane take off and leave them.

And we are ready, my craft and I. Throttles forward and outboard, gauges OK, afterburners ignite, nose-wheel steering, rudder effective, line speed, rotation speed – we are off, leaving behind only a ripping, tearing, gut noise as we split into the low black overcast, afterburner glow not even visible anymore.

Steadily we climb, turning a few degrees, easing stick forward some, trimming, climbing, then suddenly – on top! On top where the moonlight is so damn marvellously bright and the undercast appears a gently rolling snow-covered field. It's just so clear and good up here, I could fly forever. This is part of what flying is all about. I surge and strain against my harness, taking a few seconds to stretch and enjoy this privileged sight.

I've already set course to rendezvous with a tanker, to take on more fuel for my work tonight. We meet after a long cut-off turn, and I nestle under him as he flies his long, delicate boom toward my innards. A slight thump/bump, and I'm receiving. No words – all light signals. Can't even thank the boomer. We cruise silently together for several minutes. Suddenly he snatches it back, a clean break, and I'm cleared, off and away.

Now I turn east and very soon cross the fence far below. Those tanker guys will take you to hell, then come in and pull you right out again with their flying fuel trucks. Hairy work. They're grand guys.

Soon I make radio contact with another craft, a big one, a gunship, painted black and flying very low. Like the proverbial spectre, he wheels and turns just above the guns, the limestone outcropping, called karst, and the mountains – probing, searching with infra-red eyes for supply trucks headed south. He has many engines and more guns. His scanner gets something in his scope, and the pilot goes into a steep bank – right over the target. His guns flick and flash, scream and moan, long amber tongues lick the ground, the trail, the trucks. I am there to keep enemy guns off him and to help him kill trucks. Funny – he can see the trucks but not the guns till they're on him. I cannot see the trucks but pick the guns up as soon as the first rounds flash out of the muzzles.

Inside my cockpit all the lights are off or down to a dim glow, showing the instruments I need. The headset in my helmet tells me in a crackling, sometimes joking voice the information I must have: how high and how close the nearest karst, target elevation, altimeter setting, safe bailout area, guns, what the other pilot sees on the trails, where he will be when I roll in.

Then, in the blackest of black, he lets out an air-burning flare to float down and illuminate the sharp rising ground. At least then I can mentally photograph the target area. Or he might throw out a big log, a flare marker, that will fall to the ground and give off a steady glow. From that point he will tell me where to strike: 50 metres east, or 100 metres south, or, if there are two logs, hit between the two.

I push the power up now, recheck the weapons settings, gun switches, gunsight setting, airspeed, altitude – roll in! Peering, straining, leaning way forward in the harness, trying so hard to pick up the area where I know the target to be – it's so dark down there.

Sometimes when I drop, pass after pass, great fire balls will roll and boil upward and a large, rather rectangular fire will let us know we've hit another supply truck. Then we will probe with firepower all around that truck to find if there are more. Often we will touch off several, their fires outlining the trail or truck park. There are no villages or hooches for miles around; the locals have been gone for years. They silently stole away the first day those big trucks started plunging down the trails from up north. But there are gun pits down there – pits, holes, reveted sites, guns in caves, guns on the karst, guns on the hills, in the jungles, big ones, little ones.

Many times garden-hose streams of cherry balls will arc and curve up, seeming to float so slowly toward me. Those from the smaller-calibre, rapid-fire quads; and then the big stuff opens up, clip after clip of 37 mm and 57 mm follow the garden hose, which is trying to pinpoint me like a search light. Good fire discipline – no one shoots except on command.

But my lights are out, and I'm moving, jinking. The master fire controller down there tries to find me by sound. His rising shells burst harmlessly around me. The heavier stuff in clips of five and seven rounds goes off way behind.

Tonight we are lucky – no 'golden BB'. The golden BB is that one stray shell that gets you. Not always so lucky. One night we had four down in Death Valley – that's just south of Mu Gia Pass. Only got two people out the next day, and that cost a Sandy (A-1) pilot. 'And if the big guns don't get you, the black karst will,' goes the song. It is black, karsty country down there.

Soon I have no more ammunition. We, the gunship and I, gravely thank each other, and I pull up to thirty or so thousand feet, turn my navigation lights back on, and start across the Lao border to my home base. In spite of an air-conditioning system working hard enough to cool a five-room house, I'm sweating. I'm tired. My neck is sore. In fact, I'm sore all over. All those roll-ins and diving pullouts, jinking, craning your head, looking, always looking around, in the cockpit, outside, behind, left, right, up, down. But I am headed home, my aircraft is light and more responsive.

Too quickly I am in the thick, puffy thunder clouds and rain of the southwest monsoon. Wild, the psychedelic green, wiry, and twisty St Elmo's fire flows liquid and surrealistic on the canopy a few inches away. I am used to it – fascinating. It's comforting, actually, sitting snugged up in the cockpit, harness and lap belt tight, seat lowered, facing a panel of red-glowing instruments, plane buffeting slightly from the storm. Moving without conscious thought, I place the stick and rudder pedals and throttles in this or that position – not so much mechanically moving things, rather just willing the craft to do what I see should be done by what the instruments tell me.

I'm used to flying night missions now. We 'night owls' do feel rather elite, I suppose. We speak of the day pilots in somewhat condescending tones. We have a black pilot who says, 'Well, day pilots are OK, I guess, but I wouldn't want my daughter to *marry* one.' We have all kinds: quiet guys, jokey guys (the Jewish pilot with the fierce black bristly moustache who asks, 'What is a nice Jewish boy like me doing over here, killing Buddhists to make the world safe for Christianity?'), noisy guys, scared guys, whatever. But all of them do their job. I mean night after night they go out and get hammered and hosed, and yet keep right at it. And all that effort, sacrifice, blood going down the tubes. Well, these thoughts aren't going to get me home. This is not time to be thinking about anything but what I'm doing right now.

I call up some people on the ground who are sitting in darkened, black-out rooms, staring at phosphorescent screens that are their eyes to the night sky. Radar energy reflecting from me shows them where I am. I flick a switch at their command and trigger an extra burst of energy at them so they have positive identification. By radio they direct me, crisply, clearly, to a point in space and time that another man in another darkened room by a runway watches anxiously. His eyes follow a little electronic bug crawling down a radar screen between two converging lines. His voice tells me how the bug is doing, or how it should be doing. In a flat, precise voice the radar controller keeps up a constant patter – 'Turn left two degrees . . . approaching glide path . . . prepare to start descent in four miles.'

Inside the cockpit I move a few levers and feel the heavy landing gear thud into place and then counteract the nose rise as the flaps grind down. I try to follow his machine-like instructions quite accurately, as I am very near the ground now. More voice, more commands, then a glimmer of approach lights, and suddenly the wet runway is beneath me. I slip over the end, engines whistling a down note as I retard the throttles, and I'm on the ground at last.

If the runway is heavy with rain, I lower a hook to snatch a cable laid across the runway that connects to a friction device on each side. The deceleration throws me violently into my harness as I stop in less than 900 ft from nearly 175 m.p.h. And this is a gut-good feeling.

Then the slow taxi back, the easing of tension, the good feeling. Crew chiefs with lighted wands in their hands direct me where to park; they chock the wheels and signal me with a throat-cutting motion to shut down the engines. Six or seven people gather around the airplane as the engines coast off, and I unstrap and climb down, soaking wet with sweat.

'You OK? How did it go? See anything, get anything?' They want to know these things and they have a right to know. Then they ask, 'How's the airplane?' That concern always last. We confer briefly on this or that device or instrument that needs looking after. And then I tell them what I saw, what I did. They nod, grouped around, swear softly, spit once or twice. They are tough, and it pleases them to hear results.

The crew van arrives, I enter and ride through the rain – smoking a cigarette and becoming thoughtful. It's dark in there, and I need this silent time to myself before going back to the world. We arrive and, with my equipment jangling and thumping about me, I enter the squadron locker room, where there is always easy joking among those who have just come down. Those that are suiting up are quiet, serious, going over the mission brief in their minds, for once on a night strike they cannot look at maps or notes or weapon settings.

They glance at me and ask how the weather is at The Pass. Did I see any thunderstorms over the Dog's Head? They want to ask about the guns up tonight, but know I'll say how it was without their questioning. Saw some light ZPU (automatic weapons fire) at The Pass, saw someone getting hosed at Ban Karai, nothing from across the border. Nobody down, quiet night. Now all they have to worry about is thrashing through a couple of hundred miles of lousy weather, letting down on instruments and radar into the black karst country and finding their targets. Each pilot has his own thoughts on that.

Me, I'll start warming up once the lethargy of finally being

246

back from a mission drains from me. Funny how the mind/body combination works. You are all 'hypoed' just after you land, then comes a slump, then you're back up again but not as high as you were when you first landed. By now I'm ready for some hot coffee or a drink (sometimes too many), or maybe just letter writing. A lot of what you want to do depends on how the mission went.

I debrief and prepare to leave the squadron. But before I do, I look at the next days' schedule. Is it an escort? Am I leading? Where are we going? What are we carrying? My mind unrolls pictures of mosaics and gun-camera film of the area. Already I'm mechanically preparing for the next mission.

And so it goes – for a year. And I like it. But every so often, especially during your first few months, a little wisp of thought floats up from way deep in your mind when you see the schedule. 'Ah, no, not tonight,' you say to yourself. 'Tonight I'm sick – or could be sick. Just really not up to par, you know. Maybe, maybe I shouldn't go.' There's a feeling – the premonition that tonight is the night I don't come back. But you go anyhow and pretty soon you don't think about it much anymore. You just don't give a fat damn. After a while, when you've been there and see what you see, you just want to go fight! To strike back, destroy. And then sometimes you're pensive – every sense savouring each and every sight and sound and smell. Enjoying the camaraderie, the feeling of doing something. Have to watch that camaraderie thing though – don't get too close. You might lose somebody one night and that can mess up your mind. It happens, and when it does, you get all black and karsty inside your head.

I leave the squadron and walk back through the ever-present rain that's running in little rivulets down and off my poncho. The rain glistens off trees and grass and bushes, and a ripping, tearing sound upsets the balance as another black Phantom rises to pierce the clouds.

The technology of Apollo didn't just happen; it was evolved, much of it in previous U.S. programmes to evolve ICBMs (intercontinental ballistic missiles) and most notably in several post-war programmes to develop high speed rocket-powered winged aircraft. In one of these, the Bell X-1, Chuck Yeager became in 1947 the first man to fly faster than sound. In the final North American X-15 programme many of the devices and equipment that made Apollo successful began to evolve: rocket thrusters for attitude control, instrumentation for space, pressure suits for the pilots, telemetry, etc. Apollo is perhaps seen as the culmination of more than twenty-five years' work, as well as in its own right the largest scientific research programme of all time. And though Apollo proceeded smoothly enough, there were tragedies and near-tragedies with some of the earlier rocket craft.

Here X-15 Scott Crossfield describes a wild ride in which his rival Chuck Yeager achieved Mach 2·42 – then the fastest any man had ever flown, but now little more than the cruise speed of Concorde airliners – and how his rocket plane went out of control at that speed.

38

FROM: *Always Another Dawn*

BY   A. Scott Crossfield with Clay Blair, Jr
        World Publishing, Cleveland, 1960

I was king of the race track for three weeks. Then the old master, Chuck Yeager, did it again. He shattered my record, but he nearly died doing it.

I had been expecting the coup de grâce at any moment. Chuck had been scheduled to fly the X-1-A on the day after my Mach 2 flight in the Skyrocket. But when I logged Mach 2, the Air Force team pulled back and regrouped, as the military say. Yeager now had his hands full. Pete Everest describes this Air Force record-breaking in his book: 'By this time the old X-1 record had long since been broken by both Bridgeman and Crossfield, so there was no question of keeping ahead of them. Our problem now was trying to catch up.'

In early December Chuck flew the X-1-A Mach 2 and caught up. To quote Everest again: 'We had matched Bridgeman and

Crossfield even money and now we raised the bid.' Yeager would gun the X-1-A all out.

I watched these warm-ups – between my own press conferences – with more than casual interest. The Wright Brothers Memorial dinner was just a few days away. If Chuck failed, the Navy and Douglas could publicly boast a clean sweep: Carl's 85,000-foot altitude record and my Mach 2 speed record, both set in the Skyrocket. Yeager flew on December 12.

I took up a post that day on the Edwards radio circuit, to listen in on the flight from the ground. Jack Ridley and Major Arthur 'Kit' Murray flew chase. I heard them routinely chatting on the air as the mother plane bore down on the launch point at 32,000 feet. Then like the crack of a starting pistol we heard the mother-plane pilot snap to the co-pilot:

'Drop her, Danny.'

In my mind's eye I could see the X-1-A falling rapidly away from the mother plane and Yeager adroitly moving the controls. Now I knew he would be hitting the four rocket-switches at intervals, blasting skyward. In a matter of three minutes he would reach the finish line. The seconds ticked by slowly.

'Got him in sight, Kit?' It was chase Ridley speaking to chase Murray.

'No,' Murray replied. 'He's going out of sight. Too small.' That was a good sign – for Yeager.

The radio circuit was silent. There was no word from Yeager. I dragged on a cigarette thinking: it's just like him to keep everybody on the hook.

Then suddenly all hell broke loose. Something was wrong. I became aware of it when I heard Murray and Ridley shouting over the radio to Yeager.

'Chuck! Chuck! Yeager! Where are you . . .'

Then Yeager came on the air, his voice hoarse and rasping, and barely audible:

'I'm . . . I'm down . . . I'm down to 25,000 feet . . . over Tehachapi. Don't know . . . whether I can . . . get back base or not . . .'

'At 25,000 feet?' Ridley asked incredulously.

'I'm . . . I'm . . . Christ!'

'What say, Chuck?' Ridley called. 'Chuck!'

'I say . . . don't know . . . if I tore . . . anything or not . . . but, Christ!'

Yeager was obviously in serious trouble. The word flashed across the base. Emergency trucks screamed toward the flight line. Helicopters lifted off, heading for Tehachapi. We leaned over the radio speaker, hanging on each word. Race-track competition was one thing, but now a pilot's life – a *great* pilot's life – was in jeopardy. I felt helpless – almost sick.

'Chuck from Murray,' the radio crackled. 'If you can give me altitude and heading, I'll try to check from outside.' The chase pilots were trying desperately to find Yeager's tiny craft, to guide him back to base, to tell him if his wings were still in place.

'Be down at 18,000 feet. I'm about . . . be over the base at 15,000 feet in a minute,' Yeager reported.

On the ground we cheered the master on. His last radio report indicated he would make it. His voice had new confidence.

'Yes, *sir*,' Murray snapped on the radio.

We heard the routine as Chuck jettisoned and vented fuel tanks. He sounded much better. The chase closed in.

'Does everything look okay on the airplane?' Yeager called, lining up for the lake-bed landing.

There was still time to bail out if the ship was busted. But he got little help. In his eagerness Murray had lined up on the wrong airplane, a T-33 jet trainer. Quickly Murray shifted targets and gunned his engine to close on Yeager's craft, but it was too late. Yeager was already letting down, committed.

'I don't have you, Chuck,' Murray called.

'I'm on base leg,' Chuck reported. His voice sounded firm and strong. 'I'll be landing . . . in a minute.'

We heard some additional chatter and then Yeager said:

'Going to land long. I would appreciate it if you'd get out there and get . . . this thing . . . this pressure suit. I'm hurting . . . I think I busted the canopy with my head.' He landed like the pro he is.

Yeager's had been the fastest and wildest airplane ride in history. The grim details of it spread through Edwards, hurriedly passed along by tongues stammering in disbelief and admiration.

After drop, Yeager had lighted off the four X-1-A rocket barrels

250

one by one to achieve maximum speed. He pointed the X-1-A's nose toward the deep blue and at 75,000 feet he pushed over. The X-1-A, in level flight, roared to Mach 2·42, or about 1600 miles an hour, faster by a wide margin than man had ever flown before. Then in that rarefied air, at a speed the X-1-A was not designed to fly, the plane 'uncorked'. The X-1-A tumbled wildly like a 'leaf in a tempest, a cork in a flooding stream', as Everest puts it.

The X-1-A spun uncontrollably, dropping 51,000 feet in fifty-one seconds, smashing Yeager about in the cockpit. As Yeager later recalled the experience: 'The voices have no reality in this lost moment of your life. You're taking a beating now and you're badly mauled. You can see stars. Your mind is half blank, your body suddenly useless as the X-1-A begins to tumble through the sky. There is something terrible about the helplessness with which you fall. There's nothing to hold to and you have no strength. There is only your weight knocked one way and the other as the plane drops tumbling through the air. The whole inner lining of its pressurized cockpit is shattered as you're knocked around, and its skin where you touch it is still scorching hot. Then as the airplane rolls, yaws, and pitches through a ten-mile fall, you suddenly lose consciousness. You don't know what hit you or where.'

Probably no other pilot could have come through that experience alive. Much later I asked Yeager, as a matter of professional interest, exactly how he regained control of the ship. He was vague in his reply, but he said he thought that after he reached the thick atmosphere, he had deliberately put the ship into a spin.

'A spin is something I know how to get out of,' he said. 'That other business – the tumble – there is no way to figure that out.'

The Air Force squeezed in by the skin of its teeth. Yeager's new record was triumphantly announced at the Wright Brothers Memorial dinner in Washington. Yeager received many accolades. I didn't begrudge him one of them. If ever a pilot deserved praise for a job well done, it was Yeager. After that X-1-A episode, he never flew a rocket airplane again.

Men on the moon: the most fabulous journey of exploration of my lifetime, and perhaps since Columbus. The most fantastic thing about Apollo was that half the world could watch it live, in colour on their own domestic television sets.

Michael Collins was the command module pilot – the man who stayed in lunar orbit in 1969 while his two colleagues descended and made their giant leaps for mankind – on Apollo 11, the first craft actually to set men on the moon. By any standard he is a marvellous writer, and the only astronaut so far with any obvious gift for words.

In the first excerpt, he describes the launch from Florida of the colossal Saturn V rocket that carried Apollo into preliminary earth orbit. In the second excerpt, he describes how space travel has changed his perception of life.

39

FROM: *Carrying the Fire*
BY   Michael Collins
W. H. Allen, London, 1975

When Neil gets in, I give the trout to Guenter and his crew, and they frolic around a bit. Then I kick off my yellow galoshes, grab the bar inside the center hatch, and swing my legs as far as I can over to the right. After a couple of grunts and shoves, I finally manage to get my backside into the seat, with my head on a narrow rest and my legs up above me, my feet locked into titanium clamps. It's not very comfortable, especially in this suit, which is tight in the crotch, but I can put up with anything for two and a half hours – all we have left before launch. Joe is leaning over me busily, giving oxygen hoses, communications plugs, and restraining straps one last check; then he is gone. I barely have time to grab his hand before he leaves. Fred Haise is still with us. As a good back-up crewman, he was inside the CM when we got there, running some preliminary checks and certifying switch positions, and now he is down in the lower equipment bay, where we cannot reach, helping with last-minute preparations. Finally, Fred scrambles out and closes the hatch behind him. Now, hopefully, we will see no more people for eight days.

I am everlastingly thankful that I have flown once before, and that this period of waiting atop a rocket is nothing new. I am just as tense this time, but the tenseness comes mostly from an appreciation of the enormity of our undertaking rather than from the unfamiliarity of the situation. If the two effects, physical apprehension and the pressure of awesome responsibility, were added together, they might just be too much for me to handle without making some ghastly mistake. As it is, I am far from certain that we will be able to fly the mission as planned. I think we will escape with our skins, or at least I will escape with mine, but I wouldn't give better than even odds on a successful landing and return. There are just too many things that can go wrong. So far, at least, none has, and the monster beneath us is beaming its happiness to rooms full of experts. We fiddle with various switches, checking for circuit continuity, for leaks, and for proper operation of the controls for swiveling the service module engine. There is a tiny leak in the apparatus for loading liquid hydrogen into the Saturn's third stage, but the ground figures out a way to bypass the problem. As the minutes get short, there really isn't much for me to do. Fred Haise has run through a check list 417 steps long, checking every switch and control we have, and I have merely a half dozen minor chores to take care of: I must make sure that the hydrogen and oxygen supply to the three fuel cells are locked open, that the tape recorder is working, that the electrical system is well, and that the batteries are connected in such a way that they will be available to supplement the fuel cells, that we turn off unneeded communications circuits just pior to lift-off . . . all nickel-and-dime stuff. In between switch throws, I have plenty of time to think, if not daydream. Here I am, a white male, age thirty-eight, height 5 feet 11 inches, weight 165 pounds, salary $17,000 per annum, resident of a Texas suburb, with black spot on my roses, state of mind unsettled, about to be shot off to the moon. Yes, to the moon.

On Gemini 10 the launch had been a great event, for just to get into orbit meant a lot, but this time launch is only one link in a long and fragile daisy chain that encircles the moon. Our voyage has already begun because we are going to launch toward the east and thus take advantage of the earth's rotational velocity, and we are already moving toward the east at 900 miles per hour. There is a slight jolt as an access arm swings back away from the side of the spacecraft. This means that Guenter and his people have folded their tents and silently stolen away. I have to go by feel, because I can't see out at all. The two windows on my side have a cover over them, and I won't get a chance to see any sky until three minutes after lift-off, when we will already be sixty miles high. At that point we will jettison the launch escape tower, which is our means of rocketing away from an exploding booster, and with it will go the protective cover which is denying me a view. All these things I know, yet I don't feel the high excitement at the prospect of a rocket ride that I felt before Gemini 10. Neil and Buzz seem subdued too, as we go through our various checks.

There is plenty for us to do inside the CM, however, for it is one very dense, tightly designed, 12,500-pound package. The only part of this immense stack which will make the entire round trip, it is a cone eleven feet high and thirteen feet across the base. We are lying in three individual couches suspended on a joint frame, which is built to move independently of the rest of the structure, to cushion the impact of a possible emergency landing on a hard surface. In our bulky suits we touch each other at the elbows, and if we are not careful, our arms can interfere with each other as we reach for and grab various controls. Above our faces is the main instrument panel, packed with gauges and switches which must be accessible during the times we are burning one or our many engines (counting all the various little motors on the Saturn, the service module, and the command module, we have seventy-two engines – and me a single-engine pilot by inclination). There are other panels full of more switches and controls on the bulkheads to Neil's left and to my right. Then, below Buzz's feet, lies the lower equipment bay, home of the navigational equipment, the sextant, and the telescope, and the access way to the tunnel which will eventually lead to the LM, but which now points upward at empty sky. Beneath our couches there is a crawl space, where we will sleep in enclosed hammocks, and also an array of lockers containing food, clothing, and auxiliary equipment, such as the television camera. The right-hand side of the lower equipment bay is where we urinate (we defecate wherever we and our little plastic bags end up), and the left-hand side is where we store our food and prepare it, with either hot or cold water from a little

254

spout. Most controls have been arranged for our convenience, but not all – some depend upon the paths of pipes external to our compartment, such as a valve for bypassing glycol coolant fluid, which is wedged into a recess in the wall below Neil's couch and which can only be turned with a special tool. Nearly every available cubic inch of space has been used, save for two great holes in the lower equipment bay which are reserved for the boxes of moon rocks to be brought back from the LM.

The walls of the spacecraft have been decorated with tiny squares of Velcro, giving it a pockmarked look. Velcro comes in two varieties, male and female, and when pressed against each other, they cling together in a fairly secure bond. The female half consists of a felt-like, loosely woven, fuzzy surface, and these patches are cemented onto each loose item, at various strategic spots. The walls of the spacecraft are adorned with hundreds of squares of the male variety, which is a coarse material from which protrude thousands of tiny stiff fabric hooks. These hooks intermesh with the wool of the female, and together they constitute a simple and practical solution to the problem of how to keep equipment from floating away when there is no gravity to hold things in place. If I want to use a light meter, and then put it 'down' for a minute, I simply mesh the square of female Velcro on its side with the square of male Velcro adjacent to the window I am using, and it will stay there, unless my elbow knocks it loose. In addition to the Velcro, the instrument panels are adorned with tiny plastic rectangles, upon which various last-minute messages have been neatly lettered. BOIL > 50, says one, a typical reminder which says that if the adjacent radiator outlet temperature exceeds 50 degrees, the spacecraft's environmental control system will attempt to bring it back down by boiling some water. S-BAND AUX TO TAPE 90 SEC PRIOR TO DUMP says another, a memory crutch to save me some embarrassment in the operation of our tape recorder. These are personalized little notes that I have stuck on some panels. In addition, there are hundreds of other labels whose nomenclature and position are standard in all command modules. There are banks of circuit breakers, a cluster of forty-eight warning lights, two artificial horizons, and two keyboards for communicating with the computer. There are over three hundred of one type of switch alone. The paint is battleship gray, and despite the wear and tear of countless test hours, the interior appears brand spanking new.

At the moment, the most important control is over on Neil's side, just outboard of his left knee. It is the abort handle, and now it has power to it, so if Neil rotates it 30 degrees counterclockwise, three solid rockets above us will fire and yank the CM free of the service module and everything below it. It is only to be used in extremis, but I notice a horrifying thing. A large bulky pocket has been added to Neil's left suit leg, and it looks as though if he moves his leg slightly, it's going to snag on the abort handle. I quickly point this out to Neil, and he grabs the pocket and pulls it as far over toward the inside of his thigh as he can, but it still doesn't look too secure to either one of us. Jesus, I can see the headlines now: 'MOONSHOT FALLS INTO OCEAN. Mistake by crew, program officials intimate. Last transmission from Armstrong prior to leaving launch pad reportedly was "Oops."'

Inevitably, as the big moment approaches, its arrival is announced by the traditional backward count toward zero. Anesthetists and launch directors share this penchant for scaring people, for increasing the drama surrounding an event which already carries sufficient trauma to command one's entire consciousness. Why don't they just hire a husky-voiced honey to whisper, 'Sleep, my sweet' or 'It's time to go, baby'? Be that as it may, my adrenalin pump is working fine * as the monster springs to life. At nine seconds before lift-off, the five huge first-stage engines leisurely ignite, their thrust level is systematically raised to full power, and the hold-down clamps are released at T-zero. We are off! And do we know it, not just because the world is yelling 'Lift-off' in our ears, but because the seats of our pants tell us so! Trust your instruments, not your body, the modern pilot is always told, but this beast is best felt. Shake, rattle, and roll! Noise, yes, lots of it, but mostly motion, as we are thrown left and right against our straps in spasmodic little jerks. It is steering like crazy, like a nervous lady driving a wide car down a narrow alley, and I just hope it knows where it's going, because for the first ten

---

* Actually all three of us have profited from our previous space flights. The maximum heart rates we record during the Saturn's boost are: Armstrong 110 beats per minute, Collins 99, and Aldrin 88 – all considerably below the values we recorded during equivalent periods of our Gemini flights.

seconds we are perilously close to that umbilical tower. I breathe easier as the ten-second mark passes and the rocket seems to relax a bit also, as both the noise and the motion subside noticeably. All my lights and dials are in good shape, and by stealing a glance to my left, I can tell that the other two thirds of the spacecraft is also behaving itself. All three of us are very quiet – none of us seems to feel any jubilation at having left the earth, only a heightened awareness of what lies ahead. This is true of all phases of space flight: any pilot knows from ready-room fable or bitter experience that the length of the runway behind him is the most useless measurement he can take; it's what's up ahead that matters. We know we cannot dwell on those good things that have already happened, but must keep our minds ever one step ahead, especially now, when we are beginning to pick up speed. There is no sensation of speed, I don't mean that, but from a hundred hours of study and simulation, I know what is happening in the real world outside that boost protective cover, even if I can't see it. We have started slowly, at zero velocity relative to the surface of the earth, or at nine hundred miles per hour if one counts the earth's rotational velocity. But as the monster spews out its exhaust gases, Newton's second law tells us we are reacting in the opposite direction. In the first two and a half minutes of flight, four and a half *million* pounds of propellant will have been expended, causing our velocity relative to the earth to jump from zero to nine thousand feet per second, which is how we measure speed. Not miles per hour, or knots, but feet per second, which makes it even more unreal.

The G load builds slowly past 4, but no higher; unlike the Titan, the Saturn is a gentleman and will not plaster us into our couches. The 4·5 are but a little smooth, letting us know that the first-stage tanks are about empty and ready to be jettisoned. Staging, it is called, and it's always a bit of a shock, as one set of engines shuts down and another five spring into action in their place. We are jerked forward against our straps, then lowered gently again as the second stage begins its journey. This is the stage which whisperers have told us to distrust, the stage of the brittle aluminium, but it seems to be holding together, and besides, it's smooth as glass, as quiet and serene as any rocket ride can be. We are high above the disturbing forces of the atmosphere now, and the second stage is taking us on up to one hundred miles, where the third stage will take over and drive us downrange until we reach the required orbital velocity of 25,500 feet per second. At three minutes and seventeen seconds after lift-off, precisely on schedule, the launch escape rocket fires (no longer being needed). As it pulls away from our nose, it carries with it the protective cover that has been preventing me from seeing out my windows. Now it's much brighter inside the cockpit, but there is nothing to see outside but black sky, as we are already above all weather, at two hundred miles downrange from the Cape and pointed up.

As each minute passes, Houston tells us we are GO (all is well), and we confirm that everything looks good to us. At nine minutes the second stage shuts down, and briefly we are weightless, awaiting the pleasure of the third-stage engine. Due to the heightened awareness that always comes at these important moments, my sense of time is distorted, and it seems to take forever for the third stage to light. Finally! Ignition, and we are on our way again, as the single engine pushes us gently back into our couches. This third stage has a character all its own, not nearly as smooth as the second stage, but crisp and rattly. It vibrates and buzzes slightly, not alarmingly so, but with just enough authority to make me delighted when it finally shuts down on schedule at eleven minutes and forty-two seconds. 'Shutdown,' Neil says quietly, and we are in orbit, suspended gently in our straps. The world outside my window is breathtaking; in the three short years since Gemini 10, I have forgotten how beautiful it is, as clouds and sea slide majestically and silently by. We are 'upside down', in that our heads are pointed down toward the earth and our feet toward the black sky, and this is the position in which we will remain for the next two and a half hours in earth orbit, as we prepare ourselves and our machine for the next big step, the translunar injection burn which will propel us toward the moon. The reason for the heads-down attitude is to allow the sextant, in the belly of the CM, to point up at the stars, for one of the most important things I must do is take a couple of star sightings to make sure that our guidance and navigation equipment is working properly before we decide to take the plunge and leave our safe earth orbit.

Just as on Gemini 10, the first few minutes in orbit are busy ones, as a long check list must be followed to convert the spacecraft from a passive payload to an active orbiter. Between Bermuda and the Canary Islands, I work my way swiftly through a couple of pages of miscellaneous chores, opening and closing circuit breakers, throwing switches, and reading instructions for Neil and Buzz to do likewise. Then we all remove our helmets and gloves, and I fold down the bottom half of my couch and slip over it into the lower equipment bay. Here there are more switch panels, plus lockers full of equipment which I must unpack and distribute, and, of course, the all-important navigational instruments, the sextant and the telescope, which must be checked. I move slowly and cautiously, with no unnecessary head movements, for this is a phase of the flight I have been warned about. This is the first chance I will have to slosh and swirl that fluid in my inner ear, the first chance to make myself sick, and I desperately want to avoid *that*, not only on general principles, but specifically because I am the only one trained to perform the transposition and docking maneuver, which is essential to retrieving our LM from its position behind us, buried inside the top of the Saturn. Therefore, I move slowly, listening to my stomach as I go. So far so good, as I move over underneath Neil's couch and hand up to him a helmet stowage bag and a tool for turning a glycol valve. Then I check out our main oxygen pressure regulator, and unstow a couple of cameras for Buzz to use.

Buzz seems to have gotten up on the wrong side of bed this morning, or at least it seems to me he's more interested in slowing me down than in helping me get through my chores. He questions a glycol pressure reading: 'O.K., now, is that normal for the discharge pressure to zap down low and to do that? Do you think, Mike?' I reassure him, and hold out a camera in front of him. 'Buzz?' 'Yes, just a second.' I can't wait for him. 'O.K., I'll just let go of it, Buzz; it will be hanging over here in the air.' I've got to keep moving, as there are just so many minutes till TLI and so much work to be done. Let's see, the Canaries are behind us, we must be just about over our Tananarive Station on the island of Madagascar, which means Carnarvon, Australia, next; followed by the U.S.A. one last time; then around again to the middle of the

Pacific Ocean, where the TLI burn will occur. In the meantime, I have to keep things moving. 'O.K., Buzz, are you ready for the 16-millimeter?' This is the movie camera to record the transposition and docking maneuver. 'Yes, how about a bracket?' 'Neil will give you the bracket.' 'Now, let's see, you got an 18-millimeter lens on here, right?' 'Yes.' 'So – do I push the thing all the way up? Is that right?' 'Yes.' 'About with that white mark?' Lord, I don't have time to discuss camera brackets and lenses now, because it is time to take a couple of star sightings and to realign our inertial platform.

I ignore Buzz for the moment and swing around into position at my navigator's console in the middle of the lower equipment bay. I unstow and install two eyepieces, one for the sextant and one for the telescope, and I attach a portable handhold on either side of them. Handholds I need, and I let Neil and Buzz know it. 'I'm having a hell of a time maintaining my body position down here; I keep floating up.' No big problem, but annoying, as I jettison the protective covering over the optics and peer out through the telescope. What I see is disappointing, for only the brightest stars are visible through the telescope, and it is difficult to recognize them when they are not accompanied by the dimmer stars, which give each constellation its distinctive visual pattern. The situation is not helped by the fact that I am looking for Menkent and Nunki, two of the more nondescript Apollo navigation stars. Some stars are great: Antares, for instance, is not only very easy to find, just behind the head of the scorpion, but it has a distinctive reddish color as well. Menkent, on the other hand, is awfully tough to find unless the entire constellation Centaurus is clearly visible, and Nunki (in Sagittarius) is not much better. Unlike the Gemini, however, Apollo has a fancy computer tied to the optics, and now I call on it for help; it responds by swinging the sextant around until it points at where it thinks Menkent is. Aha! There it is, in plain view, and it's a simple task for me now to align cross hairs precisely on it and push a button at the instant of alignment. Now I repeat the process using Nunki, and the computer pats me on the back by flashing the information that my measurements differ from its stored star angle data by ·01 degree. It displays this information as 00001. In M.I.T.-ese, a perfect reading of 00000 is called 'five balls'. I have scored 'four balls one'. Glenn Parker, one of the Cape simulator instructors, and I have bet a cup of coffee. On this, my first measurement, Glenn doesn't think I am going to do better than four balls two, but I think I'm going to get five balls. The bet is a standoff, and after relaying this data to the ground, I add, 'And tell Glenn Parker down at the Cape that he lucked out. He doesn't owe me a cup of coffee.' Houston has no idea what I'm talking about but dutifully agrees to pass the information on.

We are over Australia now, precisely one hour after lift-off, and with the star check behind me, I can breathe a little easier. Things are going extremely well, we are precisely on schedule, and now I have time to talk about camera brackets and other trivia. In fact, we are going to attempt to send the ground some television pictures as we approach the Baja California coast, and now I unstow the TV camera, its cable, and the small monitor set which allows us to see the same picture we are transmitting. It is still dark as we pass south of Hawaii, heading toward the second dawn of this day, and the first one for me from orbit in three years. As usual, the sun comes up with a rush, and as usual my uncouth mouth records the event. 'Jesus Christ, look at that horizon! Goddamn, that's pretty, it's unreal.' Neil agrees. 'Isn't that something? Get a picture of that.' 'Ooh, sure I will. I've lost a Hasselblad. Has anyone seen a Hasselblad floating by? It couldn't have gone far, a big son-of-a-gun like that . . . I see a pen floating loose down here, too. Is anybody missing a ball-point pen?' After a lengthy search, I finally find the camera off in a corner, too late to record the brilliant flush of sunrise, but I am nonetheless pleased to have it in hand. The bulky 70-millimeter Hasselblad could turn into a dangerous projectile at the instant the rocket engine lights for the TLI burn. Just because something is 'weightless' in space doesn't mean it has lost any of its mass. It still contains the same number of molecules and, if thrown, can do just as much damage when it hits something as it would do on earth.

If I could use only one word to describe the earth as seen from the moon, I would ignore both its size and colour and search for a more elemental quality, that of fragility. The earth appears 'fragile', above all else. I don't know why, but it does. As we walk its surface, it seems solid and substantial enough, almost infinite as it extends flatly in all directions. But from space there is no hint of ruggedness to it; smooth as a billiard ball, it seems delicately poised in its circular journey around the sun, and above all it seems fragile. Once this concept of apparent earthly fragility is introduced, one questions whether it is real or imagined, and that leads inexorably to an examination of its surface. There we find things are very fragile indeed. Is the sea water clean enough to pour over your head, or is there a glaze of oil on its surface? Is the sky blue and the cloud white, or are both obscured by yellow-brown airborne filth? Is the riverbank a delight or an obscenity? The difference between a blue-and-white planet and a black-and-brown one is delicate indeed.

We rush about like busy ants, bringing immense quantities of subsurface solids, liquids, and gases up from their hiding places, and converting them into quickly discarded solids, liquids, and waste gases which lie on or just above the surface as unholy evidence of our collective insanity. The entropy of the planet, its unavailable energy, is increasing at an alarming rate; the burning of fossil fuels is an irreversible process and can only be slowed down. At the same time, the sun shines on us whether we like it or not; yet we are making but feeble efforts to focus this energy for our use. For that matter, the sun's energy, which is produced by converting hydrogen into helium, can probably be duplicated by creating our own little thermonuclear reactors here on earth, if we put every effort into the attempt. These problems and their solutions are becoming increasingly well known and I'm sure would have been recognized had there been no space program. Anyone who has viewed our planet from afar can only cry out in pain at the knowledge that the pristine blue and whiteness he can still close his eyes and see is an illusion masking an ever more senseless ugliness below. The beauty of the planet from 100,000 miles should be a goal for all of us, to help in our struggle to make it as it appears to be.

Seeing the earth from a distance has changed my perception of the solar system as well. Ever since Copernicus' theory (that the earth was a satellite of the sun, instead of vice versa) gained wide acceptance, men have considered it an irrefutable truth; yet I submit that we still cling emotionally to the pre-Copernican, or

Ptolemaic, notion that the earth is the center of everything. The sun comes up at dawn and goes down at dusk, right? Or as the radio commercial describes sunset: 'When the sun just goes away from the sky . . .' Baloney. The sun doesn't rise or fall: it doesn't move, it just sits there, and we rotate in front of it. Dawn means that we are rotating around into sight of it, while dusk means we have turned another 180 degrees and are being carried into the shadow zone. The sun never 'goes away from the sky.' It's still there sharing the same sky with us; it's simply that there is a chunk of opaque earth between us and the sun which prevents our seeing it. Everyone knows that, but I really *see* it now. No longer do I drive down a highway and wish the blinding sun would set; instead I wish we could speed up our rotation a bit and wing around into the shadows more quickly. I do not have to force myself to call this image to mind; it is there, and occasionally, I use it for other things, although admittedly I have to stretch a bit. 'What a pretty day' makes me think that it's always a pretty day somewhere; if not here, then we just happen to be standing in the wrong place. 'My watch is fast' translated into no, it's not, it's just that you should be standing farther to the east.

I'm not completely cured, though; I still say 'in' this world, instead of 'on' it, and I still think of the North Pole as being 'up' and the South Pole as 'down,' which is absurd. Give a hundred people a picture of the earth, identify the North Pole for them, and a hundred will hold the photo with the North Pole toward their head and the South Pole toward their feet. Of course, what they are really doing, if they are standing up, is pointing the South Pole at the center of the earth and, if they are standing at the equator, pointing the North Pole at some spot in the sky, which, as the earth turns, traces a circle intersecting the plane of the ecliptic at $23\frac{1}{2}$ degrees. Now why people persist in this foolishness I don't know. In my living room I have a small framed photograph showing a thin crescent against a black background. Even though the colours are wrong, people always say, 'Oh, the moon!'; but it is the earth. The earth isn't ever supposed to be a crescent, I suppose.

Finally, flying in space has changed my perception of myself. Outwardly, I seem to be the same person, and my habits are about the same. Oh, I seem to be spending money a bit more freely now, and I am inclined to put more energy into my family and less into my job, but basically I am the same guy. My wife confirms the fact. I didn't find God on the moon, nor has my life changed dramatically in any other basic way. But although I may feel I am the same person, I also feel that I am different from other people. I have been places and done things you simply would not believe. I feel like saying; I have dangled from a cord a hundred miles up, I have seen the earth eclipsed by the moon, and enjoyed it. I have seen the sun's true light, unfiltered by any planet's atmosphere. I have seen the ultimate black of infinity in a stillness undisturbed by any living thing. I have been pierced by cosmic rays on their endless journey from God's place to the limits of the universe, perhaps there to circle back on themselves and on my descendants. If Einstein's special theory is true, my travels have made me younger by a fraction of a second than if I had stayed always on the earth's surface. The molecules in my body are different, and will remain so until the seven-year biological cycle causes them to be replaced, one by one. Although I have no intention of spending the rest of my life looking backward, I do have this secret, this precious thing, that I will always carry with me. I have not been able to do these things because of any great talent I possess; rather, it has all been the roll of the dice, the same dice that cause the growth of cancer cells, or an aircraft ejection seat to work or not. In my life so far I have been very, very lucky. Even the bad things, like the surgeon's knife, have turned out to have fortunate consequences. These events confirm my native optimism, although I have seen too many promising young lives snuffed out not to know that it can happen to me. Any death seems premature, but I really believe my own will seem *less* premature, because of what I have been able to do. At what I hope is the midway point in my life (I am forty-three), my eyes have already been privileged to see more than most men see in all their years. It is perhaps a pity that my eyes have seen more than my brain has been able to assimilate or evaluate, but like the Druids at Stonehenge, I have attempted to bring order out of what I have observed, even if I have not understood it fully.

The airplane is the symbol of the new age. At the apex of the immense pyramid of mechanical progress it opens the NEW AGE, it wings its way into it. The mechanical improvements of the fierce preparatory epoch – a hundred years' blind groping to discovery – have overthrown the basis of a civilization thousands of years old.

To-day, in front of us: mechanical civilization, the reign of the NEW AGE.

The airplane, in the sky, carries our hearts above mediocre things. The airplane has given us the bird's-eye view. When the eye sees clearly, the mind makes a clear decision.

Le Corbusier, *Aircraft*, 1935.

Bibliography

Bibliography

Some other aviation anthologies:

Duke, Neville, and Lanchbery, Edward, editors, *The Crowded Sky*. London, Cassell, 1959.

Bryden, H. G., editor, *Wings*. London, Faber & Faber, 1942.

Quittenden, Enid M., compiler, *Open the Sky!* London, Pergamon, 1965.

Taylor, John W. R., editor, *Best Flying Stories*. London, Faber & Faber, 1956.

Walbank, F. Alan, editor, *Wings of War*. London, Batsford, 1942.

Raymond, Squadron Leader R., and Langdon, Squadron Leader David, editors, *Slipstream, a RAF anthology*. London, Eyre and Spottiswoode, 1946.

Jensen, Paul, editor, *The Flying Omnibus*. London, Cassell, 1953.

Glines. Lt. Col C. V. USAF, *Lighter-than-Air Flight*. NY, Franklin Watts, 1965.